A Mysterious Presence

A Mysterious Presence
Macrophotography of Plants

Photographs by Esther Bubley and Text by Percy Knauth

Workman Publishing, New York

Library of Congress Cataloging in Publication Data

Bubley, Esther.
A mysterious presence.

1. Plants, 2. Botany—Pictorial Works.
3. Photography of plants. 4. Macrophotography.
I. Knauth, Percy 1914- II. Title.
QK50.B92 1979 582'.03'0222 77-79683
ISBN 0-911104-98-4 cloth
ISBN 0-911104-99-2

Book Design: Behri Pratt Knauth

Workman Publishing Company, Inc.
1 West 39th Street
New York, New York 10018

Manufactured in the United States of America

First printing May 1979

10 9 8 7 6 5 4 3 2 1

Contents

Introduction

This is a book for all people and all seasons. More than a book about plants, it is a book about life's beginnings, and philosophers will find much to ponder in it. Botanists will find things familiar from their textbooks and experiments — cotyledon, epicotyl, and embryo are all here — but there is something more in this book. The poet's eye will see in it the exuberant dance of a new pair of shoots just emerged from the seed. The child will find the sort of images that only a child's uninhibited imagination can conjure up. The elderly will find affirmations of the changeless wisdom they have accumulated through the years. There is even a surprising and unfettered sexuality in many of these pictures, which shows that in some respects plants and people are not as far apart as most of us think. There is the drive of physical hunger, too, shown in the almost violent way a newly formed root probes the earth for the nutrients the new plant needs.

How were these pictures taken? Most of the plants were grown — either from seed or from a cutting — in a New York City apartment or on the roof three floors above busy, midtown Broadway. Avocados really began it all. When Esther Bubley, the photographer, hung a couple of avocado seeds from toothpicks with their root ends in water, she was amazed to observe the results — the fleshy caverns that opened up as the roots began to develop, the delicate, soft tissues that slowly took shape. She got lights, a tripod, and her Nikon, and went to work; and she was soon so fascinated that she decided to try the same thing with other plants. She had no idea of making a book or even of selling her pictures. She just became absorbed in what turned out to be a journey of discovery.

Avocados led to African violets, started from leaf cuttings planted in vermiculite and grown under artificial light. As they took hold, an entirely different aspect of plant growth was revealed to the photographer's astonished eye. Then came potatoes bought at the supermarket and wild onions from Central Park, picked on early morning walks.

By that time, Esther was captivated by the plants' mysterious and beautiful world. She began thinking in smaller and smaller dimensions, until she found herself on the verge of macrophotography. She put close-up lenses on the Nikon and extended the focal length still more with bellows and rings. She started cornflower, tomato, bean, pepper, cucumber, squash, and cabbage seeds, soaking them in water until they sprouted, then photographing them in the water so that they appeared to swim in a universe all their own. She spent hours arranging plants, lights, and camera until something told her, "This is it. Here is your picture." She credits the plants with helping her. "I looked through my viewfinder," she recalls, "and I pressed the button when the plants told me to press it."

That is why this book is called *A Mysterious Presence*. Esther Bubley's pictures actually became a search for the mysterious, inner presence that makes a plant the individual, living thing it is. And plants do live. That is more than evident in the pictures you will see here.

But what is most striking is an intangible yet essential quality that shines through all of these pictures; the emotion called love. Esther Bubley fell in love with plants, and she pursued that love and its attendant quest for the hidden and mysterious presence in plants until the quest was ended, the presence revealed.

Yet even the end proved to be but a beginning. For she realized that in these pictures she had been taking for so long there was a book. And so began the long process of selecting pictures and of writing the words necessary to explain them, and now that book is here.

And the plants? They grew; grew until that little Broadway apartment looked like a jungle. But not a single one of them was thrown away. Those which she could not give to friends Esther took with her on her morning walks, and planted in the park. One of these days you may find one of them there — a cabbage growing where no cabbage ever grew before, or an unexpected potato patch...

About the Photographer

To understand and enjoy her work most fully, one must know Esther Bubley. For most of her early career she was known as one of the best in the business at photographing children — a notoriously difficult task. As one gets better acquainted with her, the reasons for her success in this become apparent: she has about her a gentle quality that children immediately understand, and that lets them accept her as one of their own. Neither bossy nor preachy, she becomes part of a group of youngsters at play; unobtrusive, practically unnoticed. Once in a while a child may catch a fleeting glimpse of her aiming her camera; then Esther smiles her quick, shy smile and the child, reassured, forgets her presence.

Although she wanted to be an artist, there was an inevitability to Esther's becoming a photographer. Her first published picture, taken when she was just out of high school in Superior, Wisconsin, was of a steaming locomotive on a frigid day. The local paper gave her a prize for it. It was one of the first she took with an Argus camera she had just bought — a replacement for the Box Brownie with which she had begun taking snapshots when she was thirteen. She is still a champion of the simple camera, maintaining that the photographer's eye counts more than advanced technology. To prove this, she once did an assignment with a Kodak Hawkeye, a 1940s-vintage successor to the Box Brownie which had only two speeds and one F-stop. The pictures, of course, were just fine.

Esther's first job, which she took to save money for art school, landed her in a photo lab. When she finally did get to art school in Minneapolis, she took a course in photography. And when, around 1940, she went to New York to look for work, she found it at *Vogue* magazine — not in taking pictures of high-fashion dresses but in photographing still-life scenes for the Christmas issue. Her next job was at the National Archives in Washington, D.C., microfilming. From this menial but useful labor she was rescued by Roy Stryker, that extraordinary godfather of modern photography who, with the Farm Security Agency task force which documented the Depression years, launched the careers of so many gifted men and women. Not long after, Stryker went to New York to continue his good works, this time for the Standard Oil Company, and Esther went with him.

Since then she has taken a great many pictures for a great many publications, and has traveled throughout most of the world. She has photographed jungles and deserts, wild animals and wilderness people, artists and technocrats, circus clowns and concert pianists, in a bewildering array. None of this has changed her in any way. She still approaches photography as she always has: with an artist's eye, an artist's desire to see beyond the immediately obvious, to find the hidden beauty that gives an object, an animal, or a human being its true meaning. And it is this which makes her great.

The Food Chain

Think of the importance of plants to life on earth, and look at how tenaciously they grow wherever there is even a speck of soil in which to root! They have a huge job to do, actually the biggest job on this planet: to supply the earth with oxygen, which all creatures need in order to respire and do work.

Put simply, plants supply oxygen by purifying the air of carbon dioxide, which they absorb and process through a chemical reaction that releases oxygen. This is precisely the opposite of what animals do: they use up oxygen and release carbon dioxide. So if an imbalance in this delicate system should occur — a reduction, for example, in the number of oxygen-producing plants relative to the number of organisms and nonliving systems producing carbon dioxide — the fragile ecological equilibrium of the world would be destroyed.

Another reason why plants are indispensable to all of earth's other forms of life is that they alone can transform the radiant energy of the sun into the chemical energy that is needed by these other life forms. Solar energy, originating in huge thermonuclear reactions in that great, glowing ball that warms our world, is by far the dominant form of energy available to us — a daily flow of calories, carried in beams of light, that dwarfs all of the energy produced by humanity with all of its machines. No wonder we are looking so hard for a way to tap this fantastic source of heat and power!

Plants can tap it right now, and have been doing so for millions of years. They can not only transform the radiant energy of the sun into usable chemical energy, but can also store the chemical energy for use when they need it, much as we store electrical energy in a storage battery. Animals eating plants also use this stored energy to fuel their bodies and do work. Other animals that kill and eat the plant-eating animals benefit from the energy that the plant-eaters derived from the plants. And even when the largest animal in the food chain falls dead in the forest, bacteria and fungi feed on its decaying body. Thus they derive from it the last bit of energy which originally came from the sun, was transformed to chemical energy by plants, and was so made available on earth.

The system by which life-sustaining nutrients are passed from one living thing to another is called a *food chain*. It leads from plants through a variety of animals and down to the lowly bacteria — and food chains involve every living thing on earth. Some food chains are very short, some much longer, but all begin with plants, and without plants, there would be none. Without plants, our planet would be nothing but a barren, sterile rock spinning coldly, like a lifeless star, through the emptiness of space.

Avocados

As it is in almost all plants, the first sign of growth in this avocado is a tiny rootlet (above). Pale and delicate, the rootlet (right) pokes its way with amazing strength past the folds and crevices of the endosperm which nourishes it, ready to take up the tough job of penetrating the harsh, resistant soil to find water and dissolved nutrient minerals.

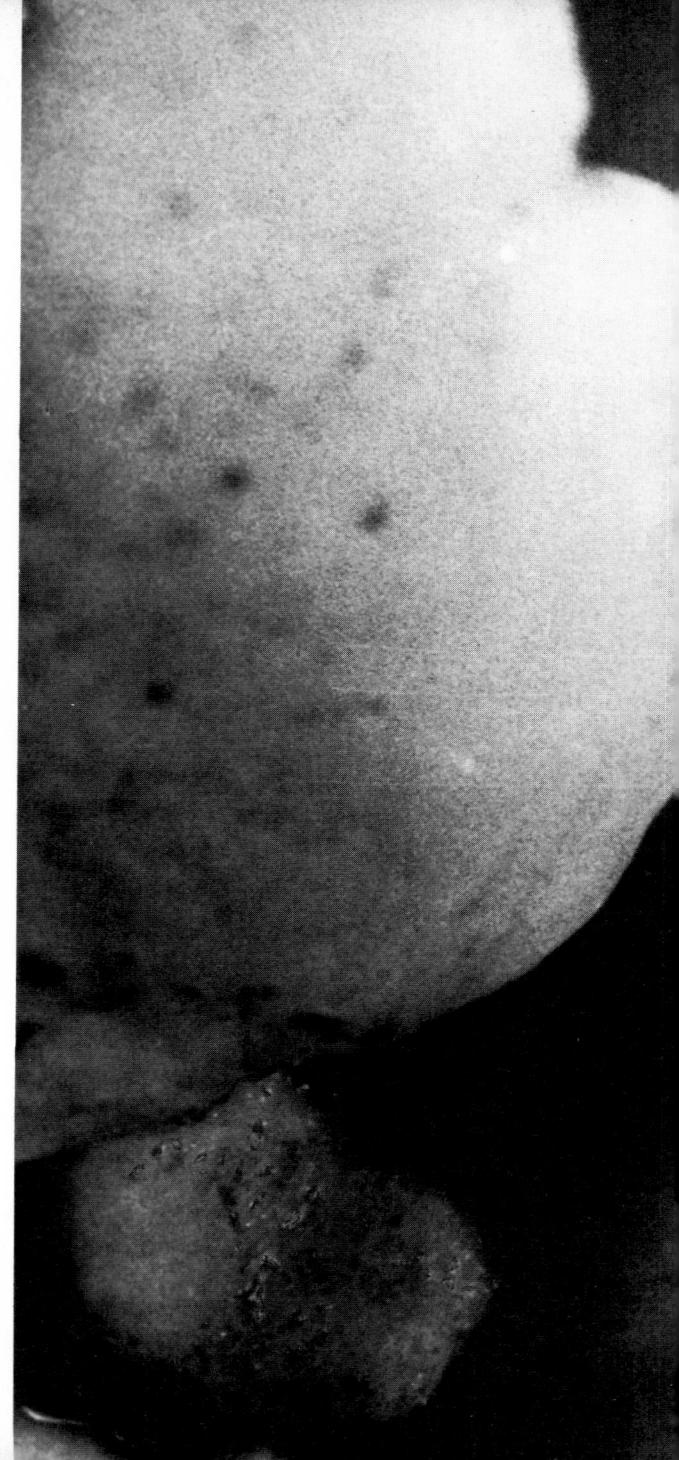

In a profusion of growth, the roots burst through the husk of the parent seed. The long root stretching across these pages beautifully illustrates how roots grow: at its whitish tip is the apical meristem, where cells, constantly dividing, push the root forward; in the darker area behind the tip, older cells are elongating, hardening, and forming the tough skins that will protect them from abrasive soil particles. Just outside the husk is an area of differentiation, where root hairs are being formed by variant cells.

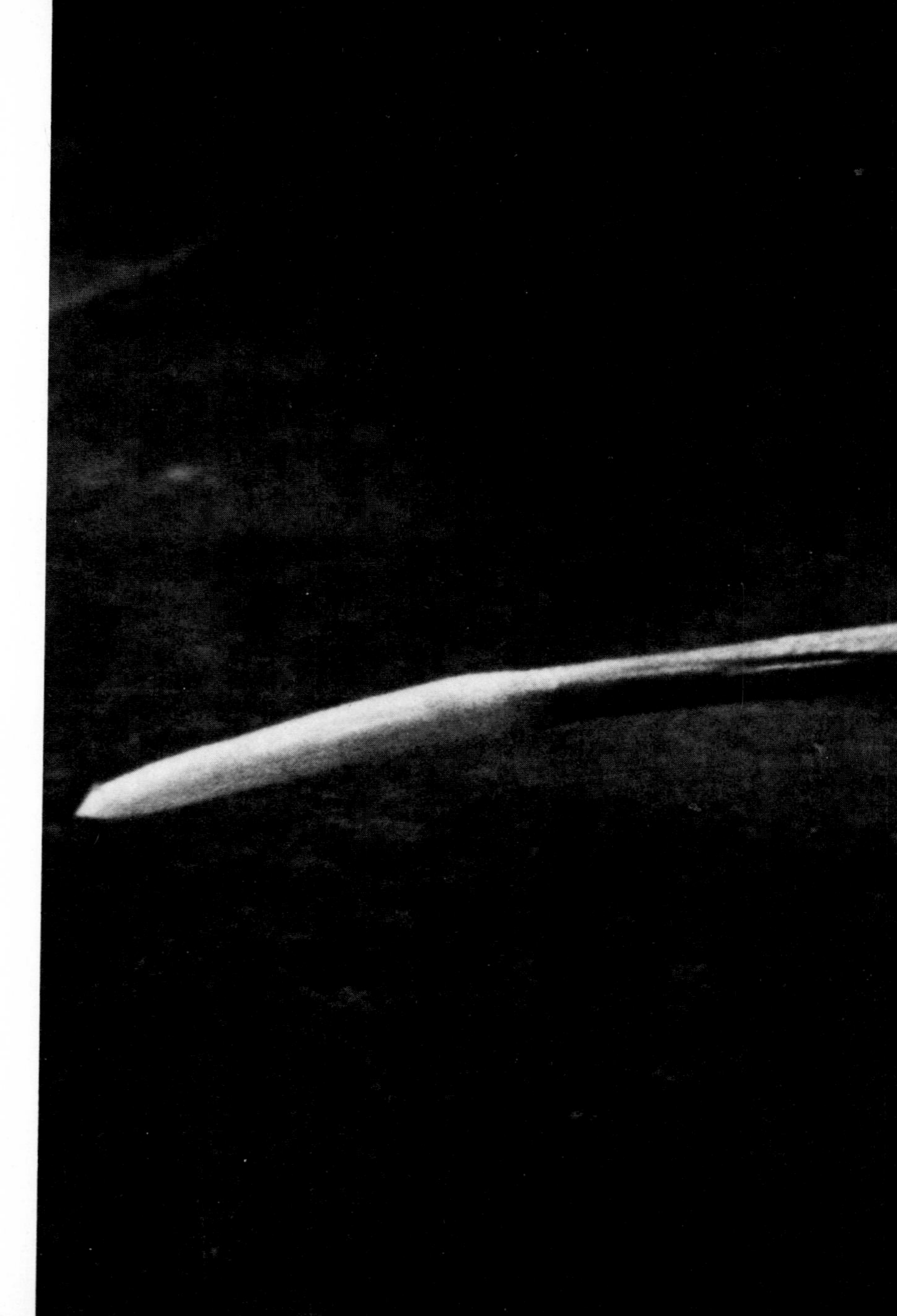

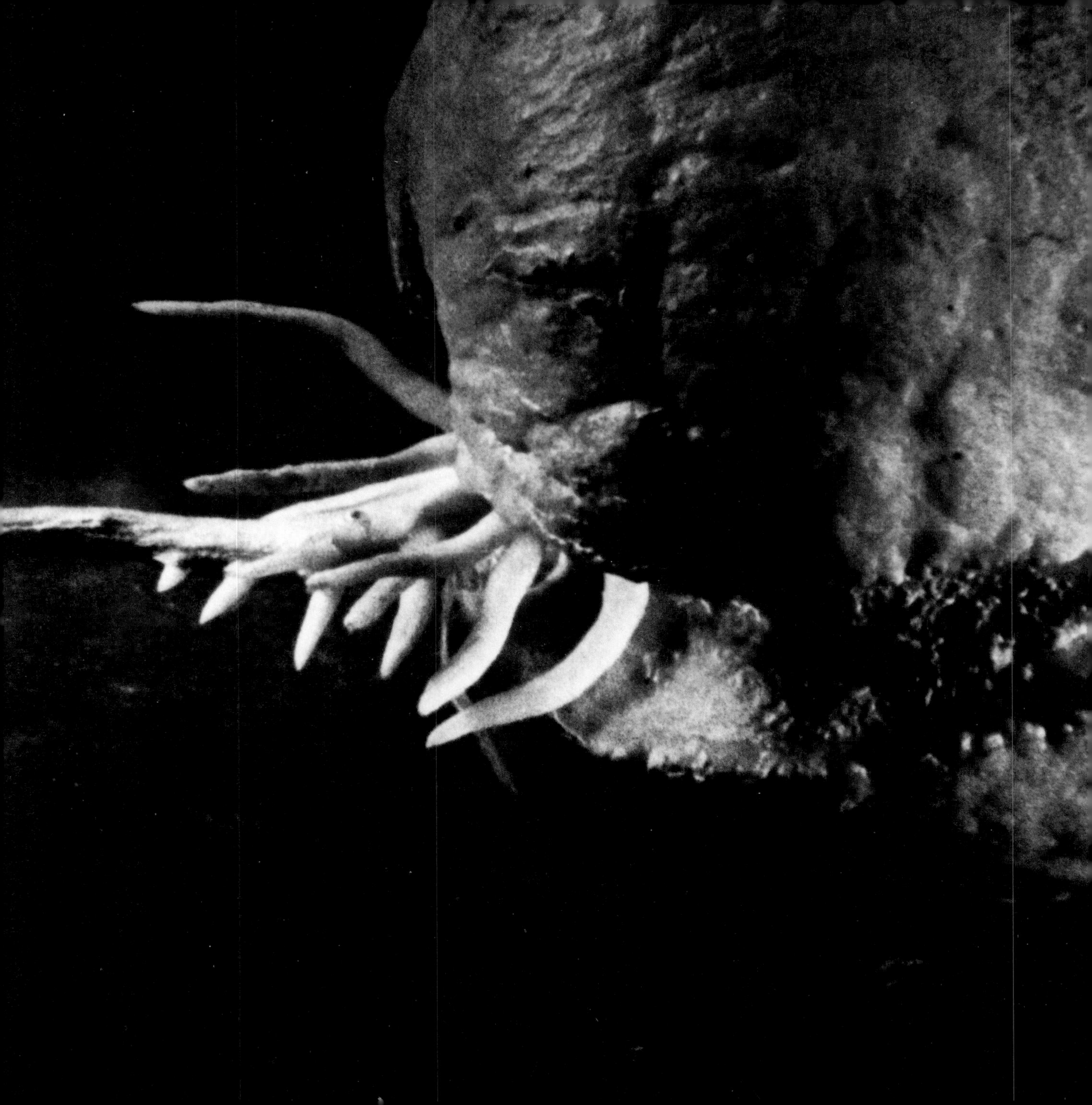

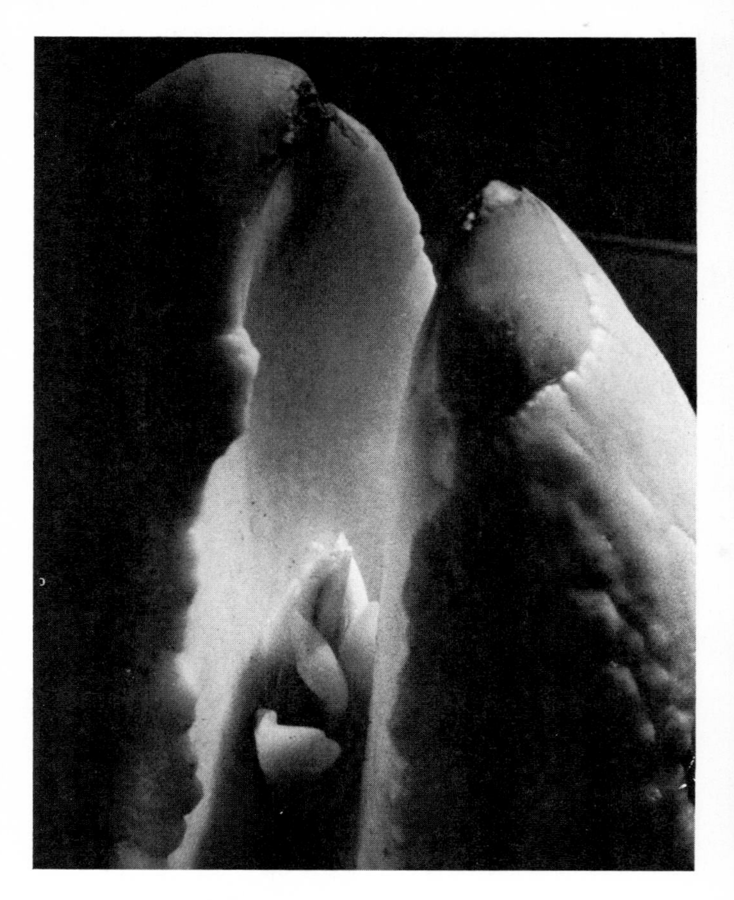

In the cleft of the splitting pit, the tip of the newborn shoot appears (above), pushing outward toward the life-giving sunlight. This is a close-up of the apical meristem, the primary area of cell growth and the most delicate part of the plant.

Pushing still further outward as the pit opens, the shoot reveals the twin protective sections of the cotyledons — the halves of the pit — which still encase and nourish it in this initial stage of its upward journey. Later, the cotyledon leaves will fall away.

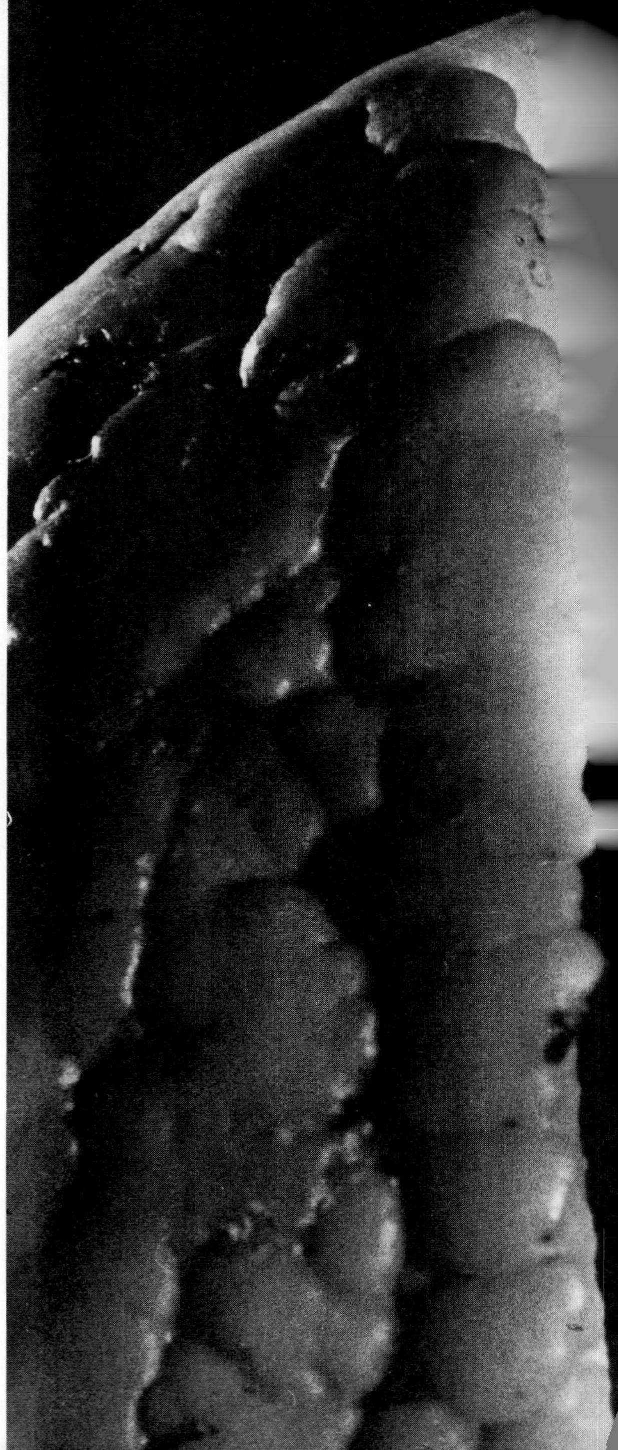

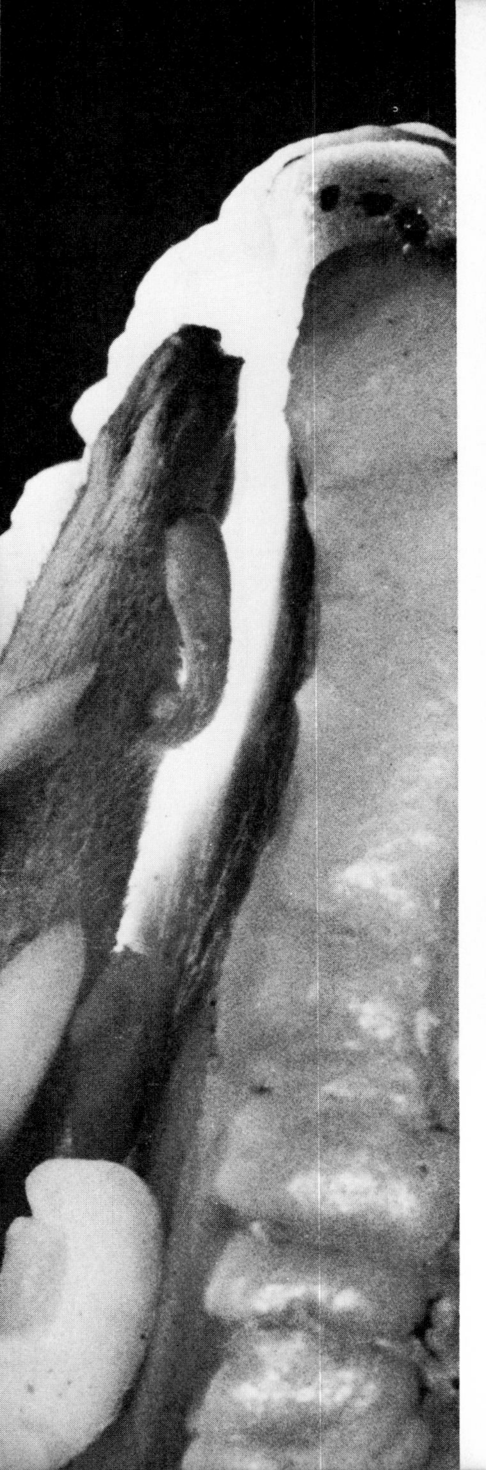

A leaf primordium (above) begins to form on the growing stem. This is the beginning of the large, umbrella-like leaves which carry out the most vital function of the plant: the manufacture, through photosynthesis, of chemical energy from the radiant energy of the sun.

Seen here in silhouette is an avocado plant in the making. The two huge black shadows on either side are the two halves of the pit, cleft by the developing embryo or new plant as it grows. The roots, at bottom, are now growing downward; the shoots have begun their journey upward. At this stage, both are still nourished by the stored food of the endosperm surrounding them, contained in the cotyledons (in this case the pit). On the opposite page, a shoot has emerged into the light above the split halves of the pit.

Against the background of the inner wall of the cloven avocado pit, a group of half a dozen shoots fights for living space, crowding up toward the light. Left to themselves, they would not produce a healthy tree; at this stage a dedicated avocado grower would prune all but the central shoot, which is clearly the strongest.

One wonders which of these shoots in various stages of development is most likely to outlast the others. In nature, following Darwin's law of survival of the fittest, the weaker shoots eventually give way to the strongest.

The leaf has finally unfurled from the bud, revealing its many delicate veins — part of the complex plumbing system that will carry sugar down to the roots. The cells of these leaves contain the marvelous mechanism which no one has ever been able to duplicate: the "chemical factory" that splits apart molecules of water and transforms the radiant energy of the sun into chemical energy the plant can use.

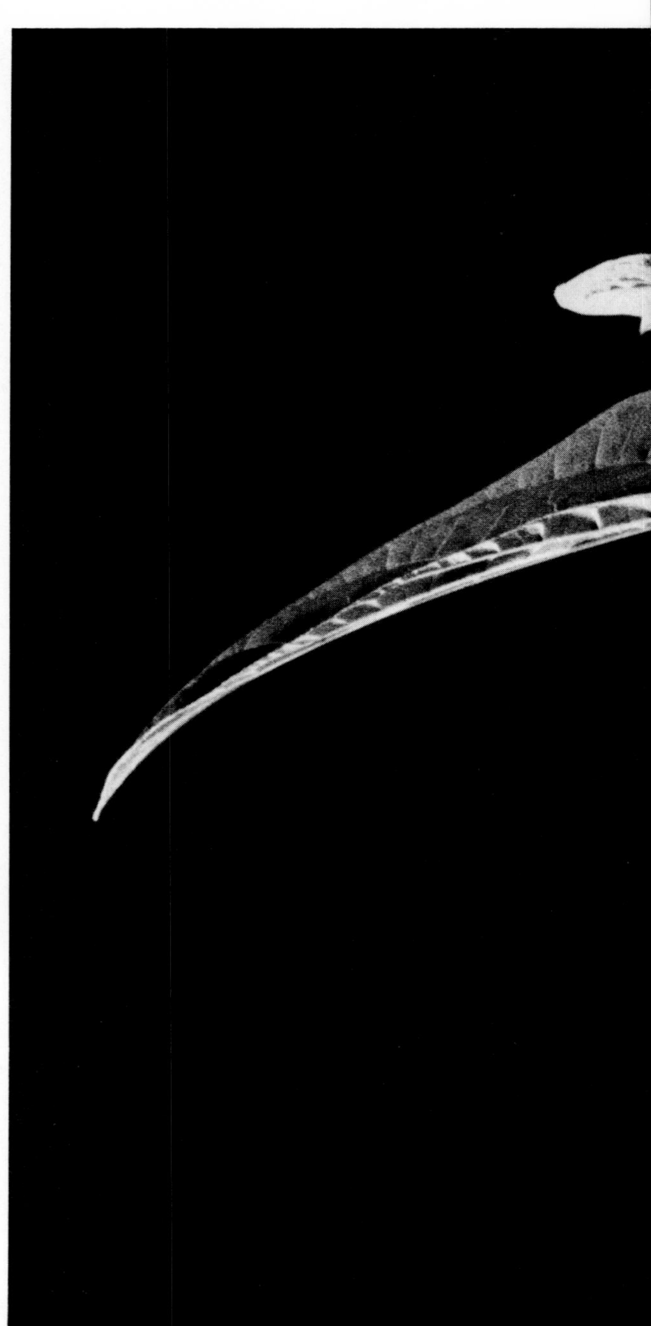

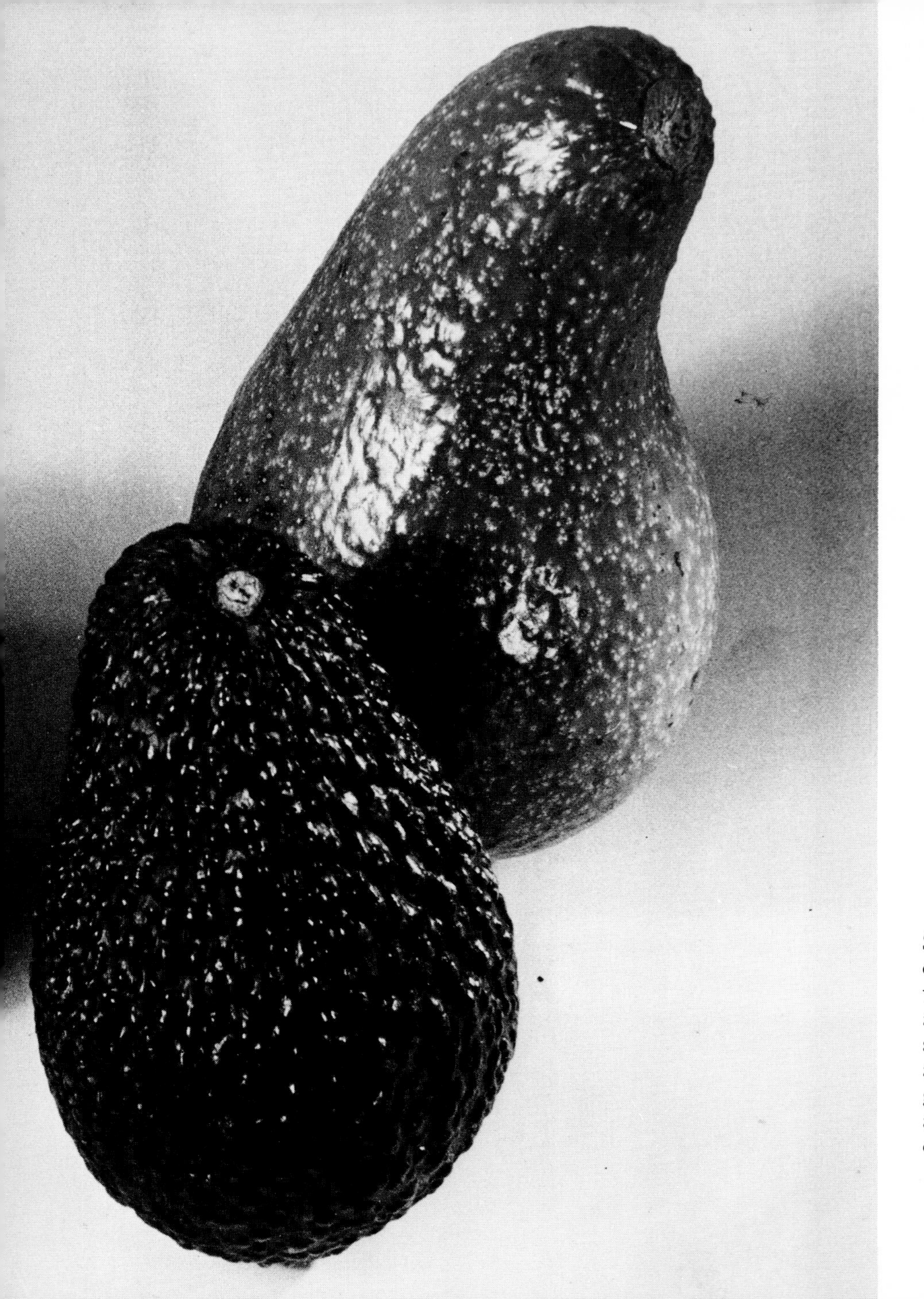

Stripped of the edible covering of soft, buttery flesh that lies within its outer husk, the avocado is tough enough to survive until it finds the warmth and wetness its seed needs to germinate. It is a tropical plant, and so, in northern climates, it can only be grown indoors.

Fertilization & Pollination

The miracle of life is dramatically illustrated in the fertilization that gives rise to a plant. It is a process at once absorbing and astonishing, because, despite the haphazard ways through which the sperm must pass to finally meet the egg cells it fertilizes, it very rarely misses. Pine trees and other conifers accomplish the meeting by immense overproduction of the yellow sperm cells — their pollen — which, when spread by the wind, literally coat trees and lakes in early summer. Flowering plants effect pollination by attracting insects which, in passing from flower to flower, distribute some of the pollen left on their bodies by the flowers they have visited. But no matter how random the method of pollination may seem, the result is always fertilization; and the consequence of this is always a new plant — not just any plant, but one typical of the species involved.

To understand the pollination process we must first picture in our minds some of the details of a typical flower. Forget the petals for a moment. There is a small central stalk — the *style* — which resembles a column and grows right up the middle of the flower from a swollen base. This base is called the *ovary*, because it contains the egg cells. The top of the style, called the *stigma*, is about even with the tips of the petals, and widens out slightly to form a rough-textured platform covered with a sticky, glue-like substance.

If a grain of pollen comes floating along on the breeze, it may brush the stigma and stick to it, trapped by the glue. If an insect comes seeking nectar from the flower, the pollen left on its legs and body from an earlier visit to another flower may brush against the stigma and get caught, too. Some flowers, like the violet, are cross-pollinated by both the wind and insects, and are also self-pollinating, in which case the pollen produced by a flower drops by itself onto the stigma of the same flower, usually before the flower opens. Peas are entirely self-pollinating, which is why the pioneer geneticist Gregor Mendel used them in formulating his epic genetic laws:

he was able to control their pollination exactly and to pollinate the flowers as he chose.

Once a grain of pollen arrives on the platform at the top of the style, it begins to grow a very thin, threadlike tube, which grows right down through the center of the style. The development of this tube is an extraordinary thing, considering the minuscule size of the pollen grain itself. The tube is a little pipe about a thousandth of an inch in diameter, and it penetrates all the way to the bottom of the style, where it enters the ovary.

In most cases, the distance through which the pollen tube grows is not too great, but in some cases it is enormous. To fertilize an ear of corn, for instance, the pollen tube must grow through all of the corn silk and then through the entire length of the ear, sometimes as much as a foot.

When the pollen tube has started to grow, the two nuclei enclosed within the pollen grain — which is also known as the *male gametophyte* — move into the tube. One of these nuclei then divides, forming two *sperm nuclei*. When the tube has penetrated the ovary, the two sperm nuclei travel down it and enter the ovary. There, each of the sperm nuclei performs a separate function. One fertilizes an egg nucleus, to create the embryo of a new plant. The other sperm nucleus unites with the *polar nuclei* — nuclei which the female gametophyte produces at the same time as it produces the egg nucleus, but which do not take part in formation of a new plant embryo. The joining of the second sperm nucleus with the polar nuclei forms the tissue for the endosperm, which will store food for the new plant embryo.

Once it is fertilized, the egg cell matures into a seed. In the flowering plants, this seed remains enclosed in the ovary while the ovary — and the nutritious endosperm tissue it contains — ripens into a fruit. In the process, the ovary grows much larger and in many cases develops a sweet, fleshy, or juicy tissue which is attractive to animals. Thus, animals eat the fruits of plants and distribute the seeds by carrying them in their intestines for a while before eliminating them somewhere else — another example of the ingenious systems that help plants to reproduce and extend their latitude.

Beets & Radishes

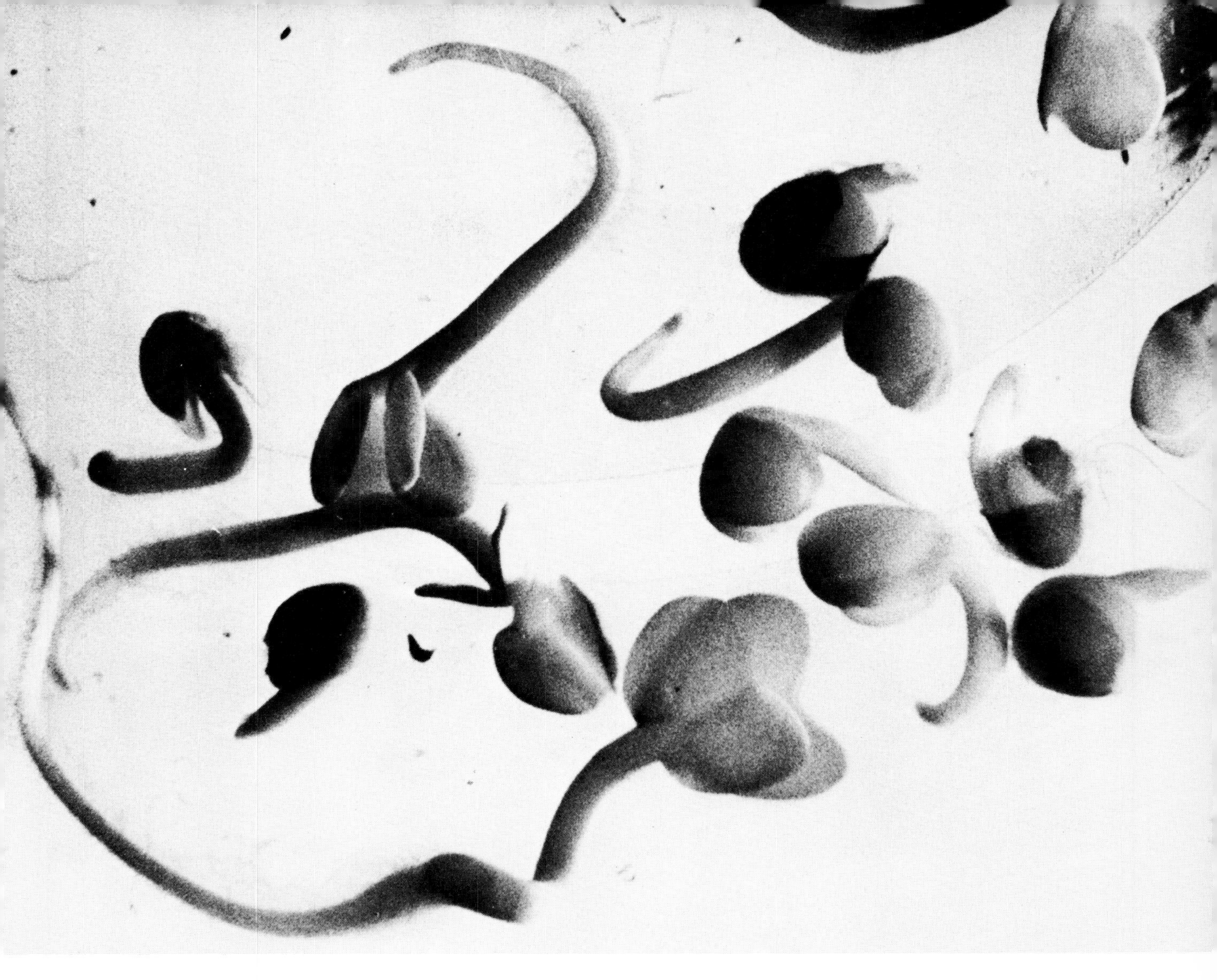

The fuzz growing on these incipient radish roots is actually root hairs. Each hair is the extension of a single cell in the outer skin of the root, and serves to convey vital nutrients to the plant.

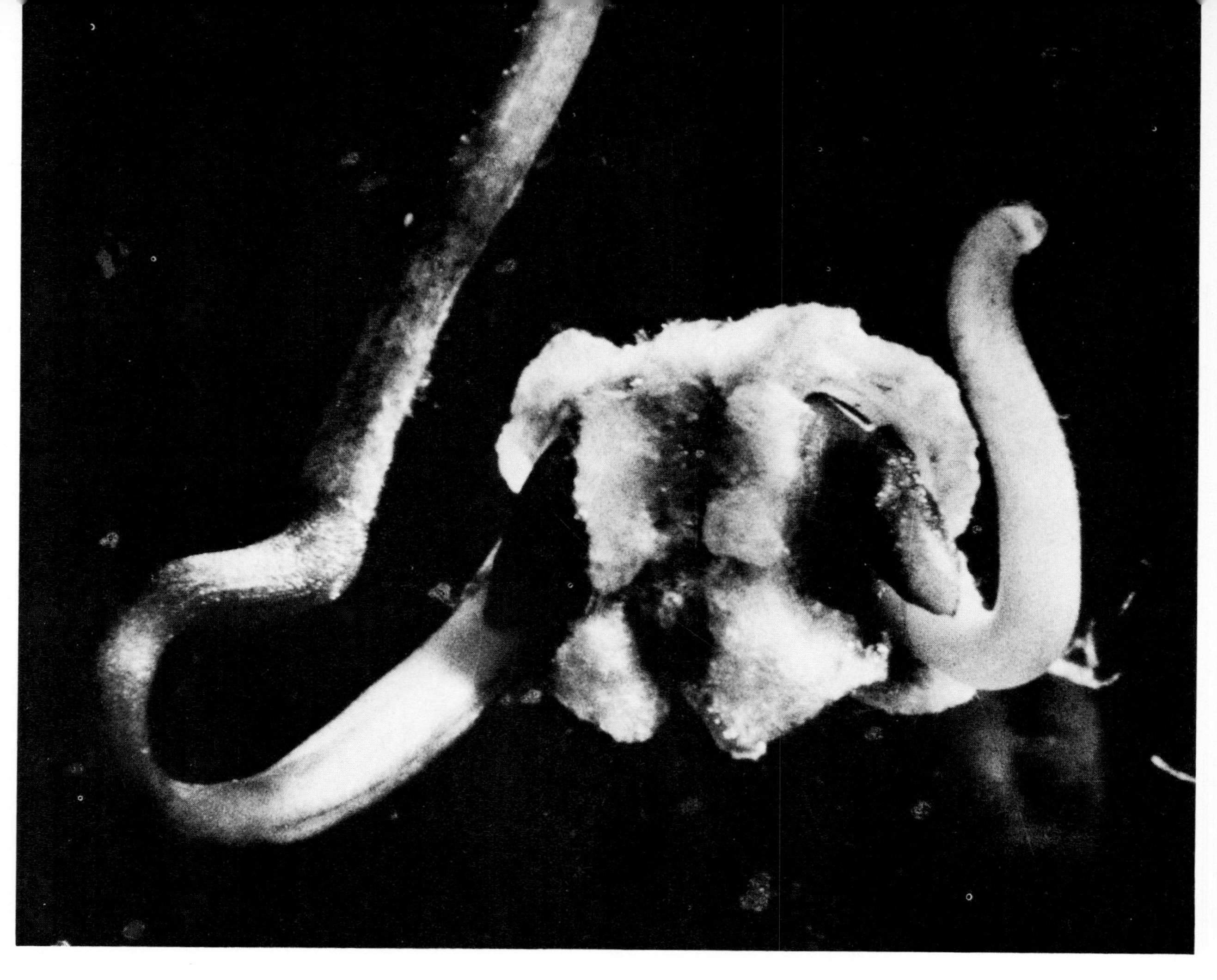

The beet seedling is just emerging from its sheltering cotyledons. Its root will soon begin to thicken and put out secondary roots. Like the radish, the beet has a taproot system, with one dominant root growing straight down, and the secondary roots branching from it. The beet uses its main root to store sugar and other nutrients, and it is this root which humans and animals eat.

The thick, fleshy stems and spreading leaves of the maturing beet plant put a heavy demand on the root system to keep them supplied with water and minerals. Here in the leaves the work of photosynthesis takes place; the sugar thus produced is then carried back to the roots where it is stored.

These beet leaves illustrate beautifully the ribs in the leaves which are part of the plant's vital transport system.

The Seed

The most intriguing miracles in the world around us are often those we never notice because they are so common. Thus we pass without thought a freshly-sown lawn, noting only with a touch of pleasure its mist of pale green covering the earth. If only we were to pause, bend down, and look to see what makes that gauzy haze! Here are thousands upon thousands of green plants, each pushing a tiny shoot bravely upward through the coarse brown soil. Here, in this little patch of freshly dug and seeded earth, are all sorts of miracles taking place.

What makes a seed grow? How can an entire plant develop from just one tiny seed? How does the plant know which way to grow? Supposing the seed falls on the ground upside down, will the plant start growing upside down? And what, after all, is a seed?

The seed is not, as we might suppose, the beginning of a plant. Rather, it represents a phase in a plant's development which we can call the "moving and storage phase." The seed is essentially a package containing all of the necessary elements for the development of a plant, securely wrapped in a tough, weatherproof skin that can withstand the onslaught of wind, cold, and prolonged immersion in water, and even the ravages of some creatures' digestive systems, so that, at the proper time, and in a hospitable place and climate, the seed skin will burst open and let the plant be born.

The story of the seed begins when an egg nucleus is fertilized by a sperm nucleus inside the *ovule* — a special chamber, several of which make up the ovary of a plant. The fusion of a second sperm nucleus with the polar nuclei — also inside the ovule — will produce the *endosperm* that will surround the growing embryo and provide food for it. The endosperm is often the part of a seed we like to eat — the inside of a sunflower seed, for instance, or the meat of a nut.

The union of male and female cells in the ovary of a plant creates what is known as a zygote. This is the true beginning of a plant, from which the embryo develops. It does so by means of specialized cell division; and the same sort of specialized cell division also results in formation of the endosperm. Finally, the ovule — containing both the embryo and the endosperm tissue that will nourish it — begins a process of thickening and hardening its walls, finally evolving into the seed — a self-contained plant ready to start growing and able to survive whatever is to come.

How a seed knows when to *germinate* — to let the plant inside it begin to grow — is a beautiful and in many ways still mysterious story. Some seeds simply will not germinate, no matter how ideal the conditions, unless they have first been through a period of prolonged cold; these, as you might guess, are northern plants. Other plants need warm, drenching rains. On the pages that follow you will see various aspects of sprouting seeds; on other pages, you will see the same processes, but they will look quite different — because there is as much difference between the myriad kinds of plants as there is between the various species of animals.

And yet, for all of the huge diversity (some 375,000 species) of plants, there is a unity to the basic process of plant development that is at once impressive and surprising. But what is even more intriguing — and here the scientist and the poet may part company for a while — is that plants have something which science cannot pin down, but which the poet's inner eye can sense: a mysterious inner presence, something very like a soul.

Nobody can keep and care for plants for any length of time without feeling this inner presence to some degree. Perhaps you can find traces of it in these pages. It may be winsome or majestic — the soul of a violet will, after all, be different from that of a Norway pine. But it will be there. And as you pursue it through this book, you may even find something else, something hauntingly familiar, like an ancient memory: a sense of something shared when the earth was very young and we were all just beginning, long, long ago.

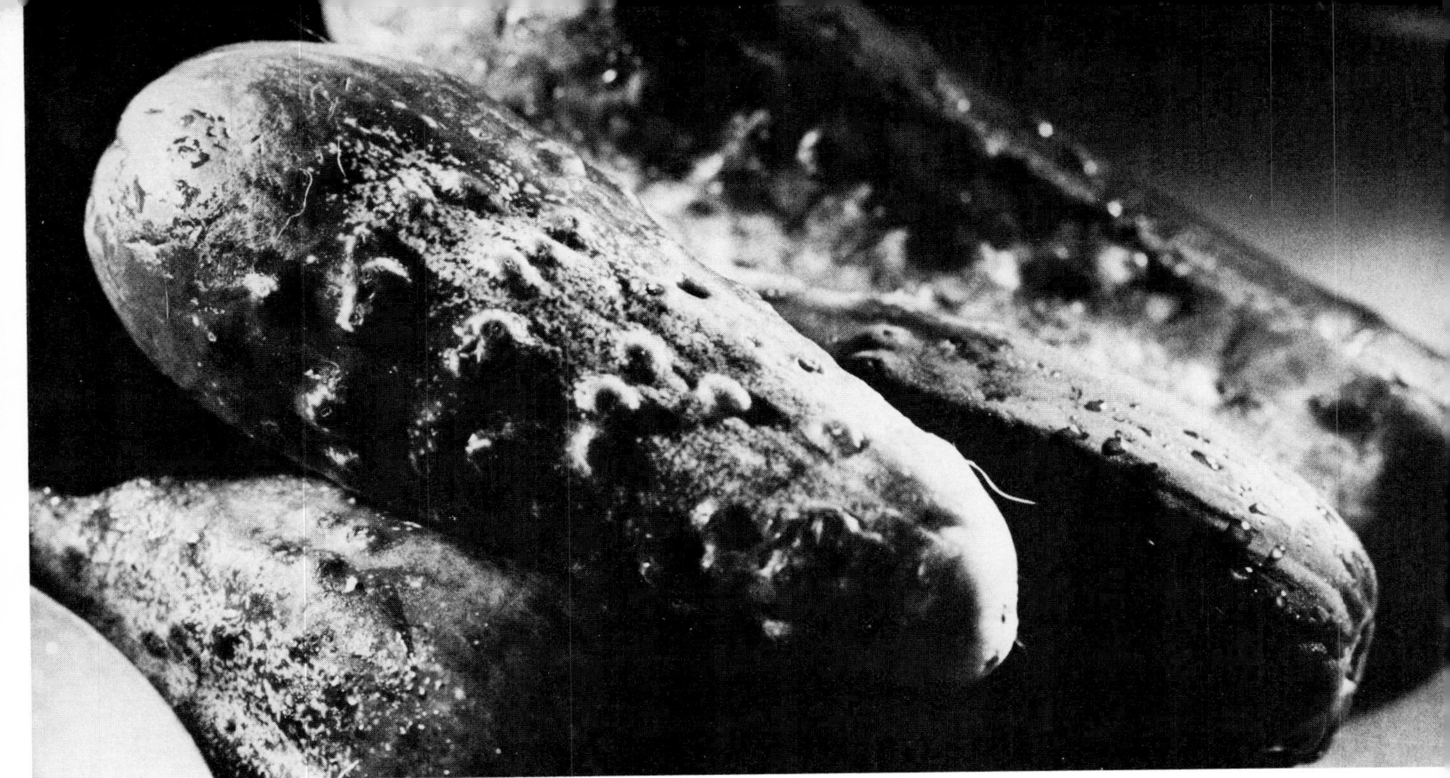

Cucumbers

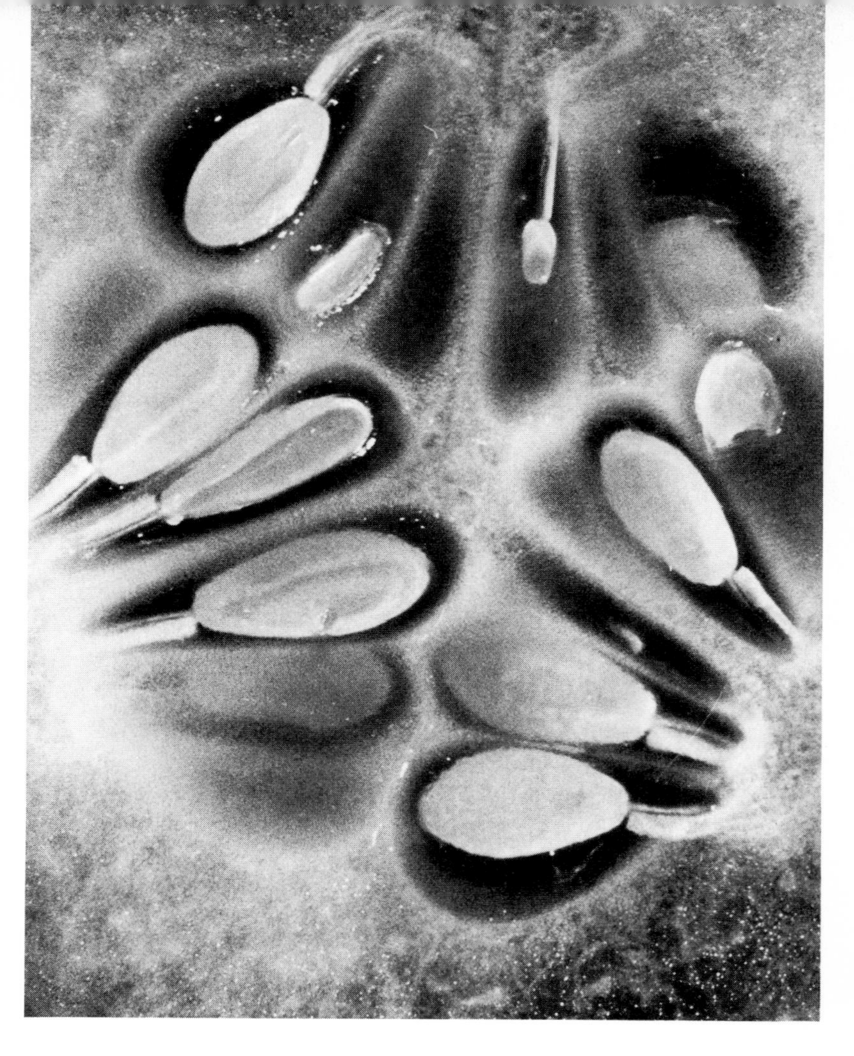

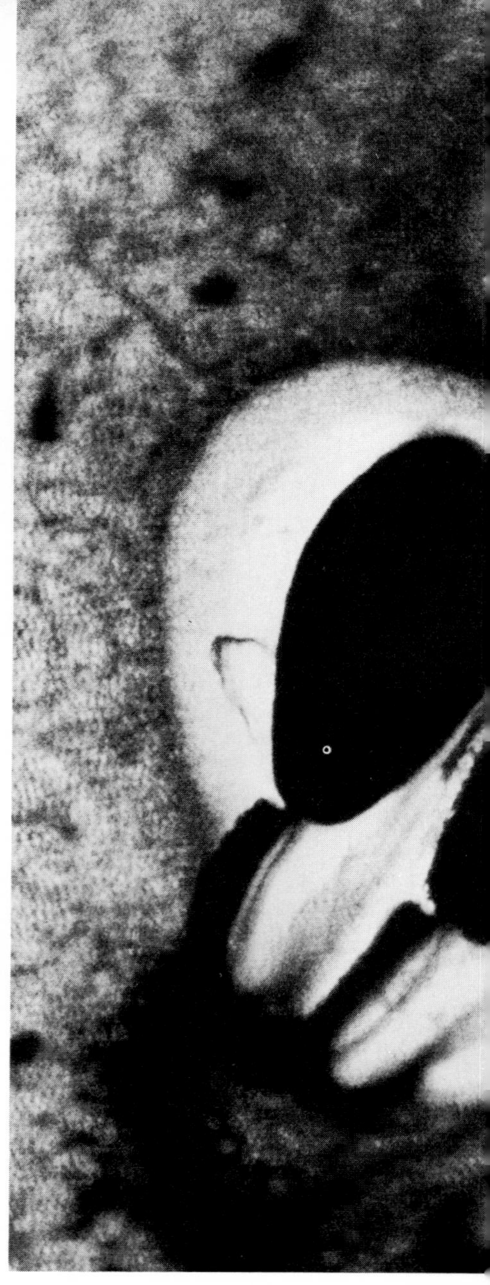

These are the seeds of a cucumber, suspended in water, surrounded by the succulent flesh of the ovary. They wait to be released by their inhibitor — be it hormonal, chemical, or light-related — so that they can germinate and grow into new plants on their own.

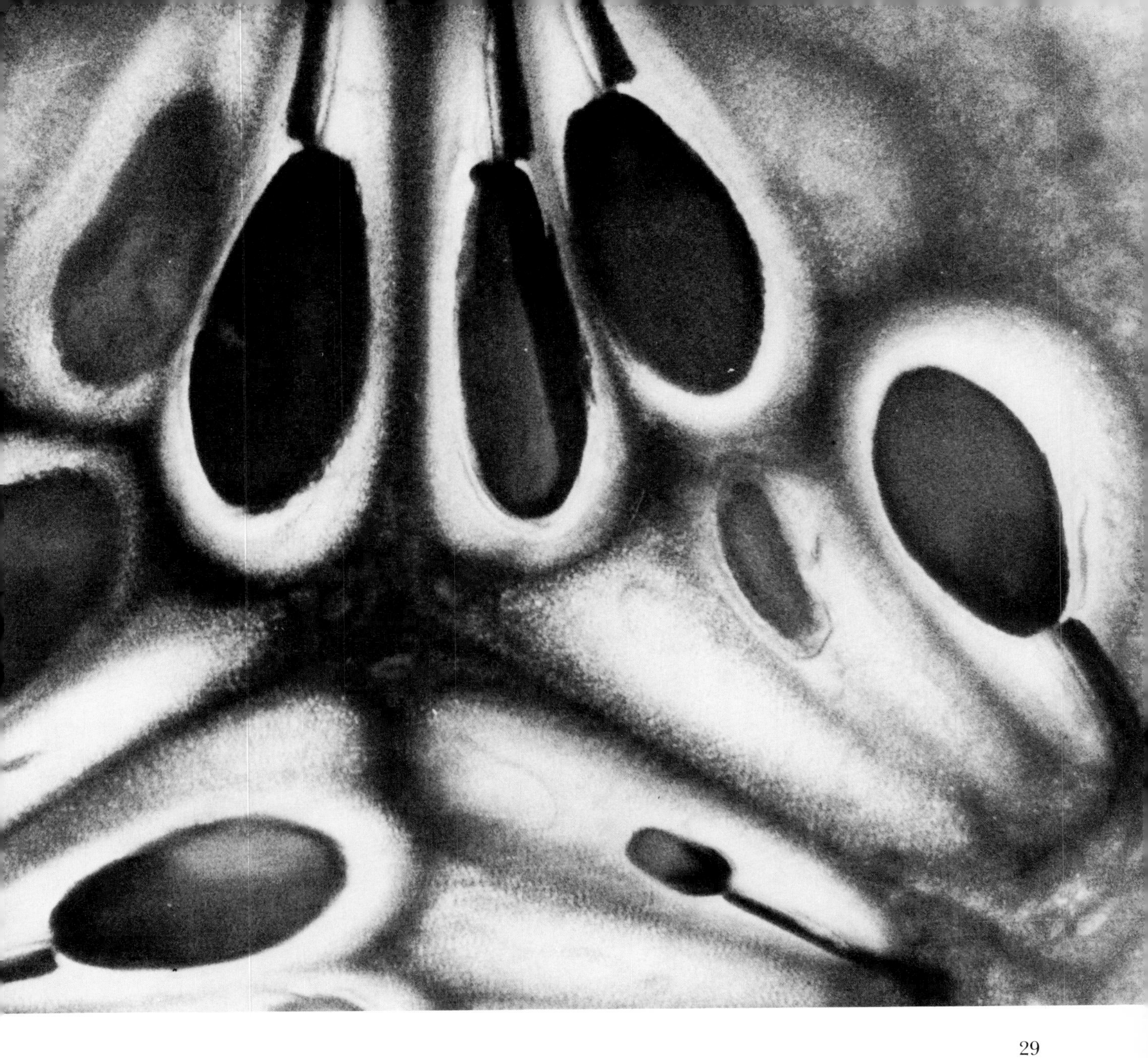

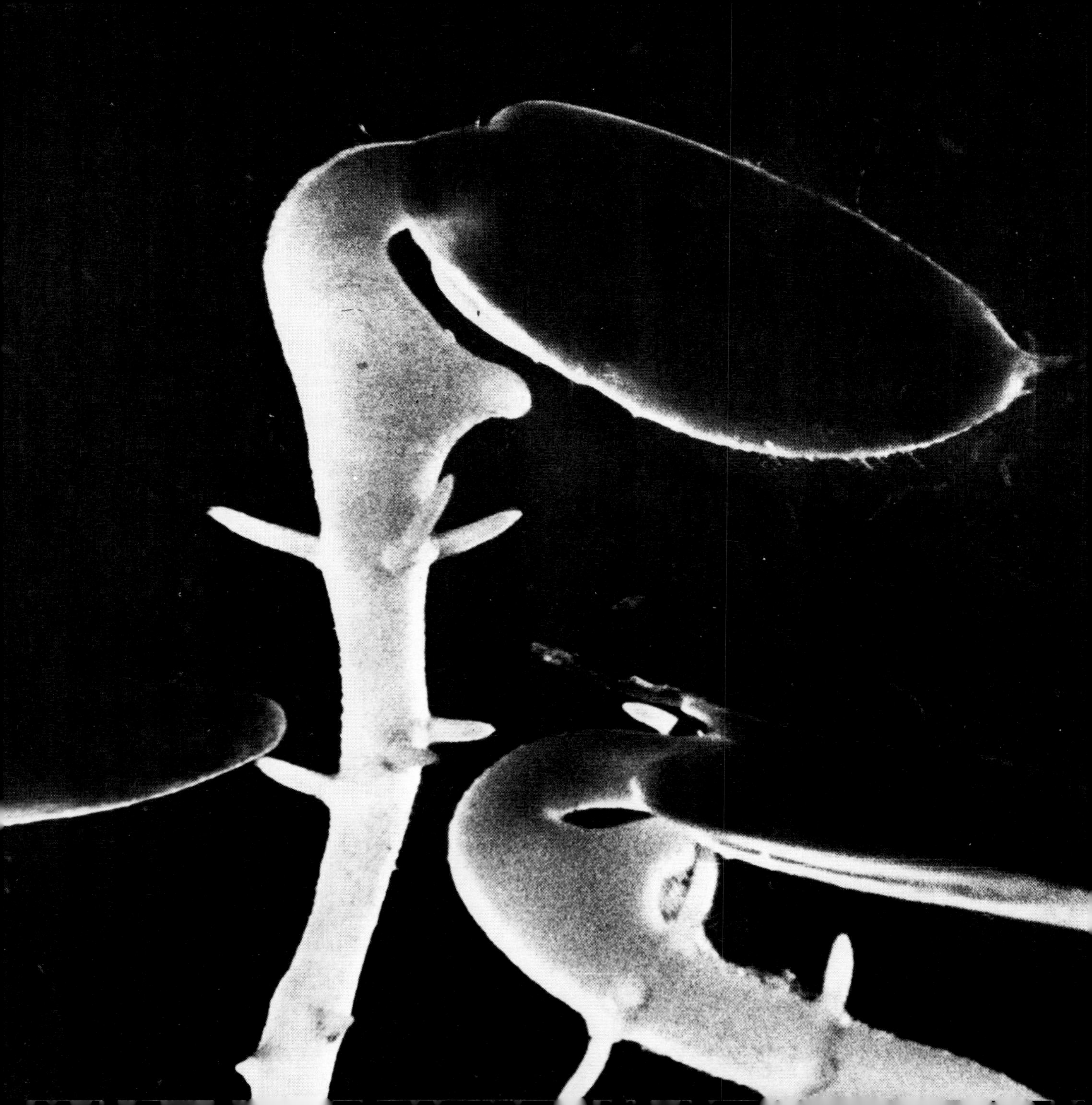

A leaflet has emerged on the left of this shoot, and the seed coat will soon be shucked off by the cotyledons whose outlines are visible within.

The process of germination advances. From one end of the seed the root shoot emerges, and tiny, spike-like lateral roots have already begun to spring from it.

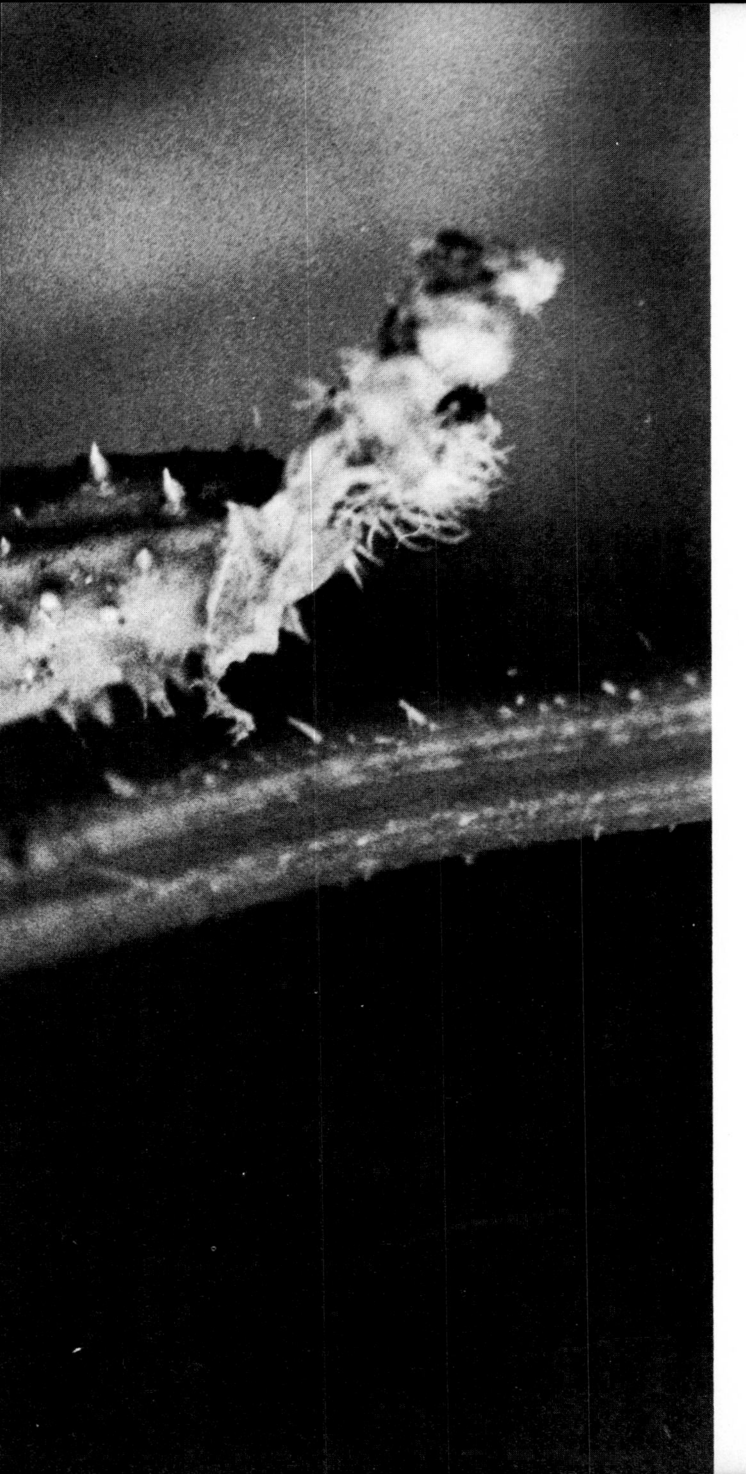

Like a caterpillar, a cucumber begins life as a hairy, rather repulsive creature. Its warts have spines on them — a measure of protection. Its skin is tough, for the same reason: it must survive as a seed case before it can be the progenitor of new cucumbers. Even that curly shoot near its tail has a function: a cucumber is a climbing plant, and tendrils like this fasten onto trees, fenceposts, or trellises to pull it up into the air.

This cucumber (right), growing upward from the axil, or juncture, of two branches is anything but appetizing; with all those spines and needle-like hairs, it looks more like a thistle than like a cucumber. At the very top is the not yet developed female flower. In the case of the cucumber, the ovary is not enclosed within the flower, but is below it. On the opposite page, growth progresses and tendrils reach out to grasp support for the growing vine.

A cucumber blossom, pale yellow in color, attracts insects for pollination. Peeping out of the center between the petals is the stigma, the organ which receives the pollen and thereby starts the process of fertilization; its tiny hairs catch the pollen grains.

Germination

Of all the many adaptations that plants have evolved through the eons in order to survive in the various environments of earth, few are more interesting or more important than those involved in the germination of the seed. This is the beginning of the life of a new plant, the time when an embryo begins its growth, when the root and the shoot form and the plant takes on an independent existence. And one of the most important aspects of germination is timing: it is as important for a seed to know when *not* to germinate as it is for it to know when the time for germination is right. Through the ages, plants have evolved some very ingenious mechanisms for just this purpose.

Consider an apple tree. By late September in the temperate zone, its ovaries will have enlarged into fruits, and the fruits, ripened to a tempting red, are ready to drop to the ground. By late October, whatever fruits have not been picked by humans or eaten by animals will have rotted — the fleshy endosperm tissue of the apple decaying — so that the seeds in the core come in contact with the earth.

Now let us suppose we get a late October rainstorm, followed by a period of warm Indian summer. All of the preconditions for germination will be present: an ample supply of water, warm temperatures, sunlight, and contact with a receptive earth. Why does the seed not germinate now?

If it did, the young plant would be doomed; for no matter how well it took hold in the earth, it would never survive the winter. If most of the earth's apple trees began reproducing in October, there would soon be very few of them left. Those which remained would be mutants which for some reason began growing in the spring. That is exactly what happened eons ago, when apple trees were first evolving, and it demonstrates the selective hand of evolution: the trees that germinated in October died, those that germinated in the spring survived.

There are a number of ways by which germination can be prevented, even when everything is in its favor. For one thing, the embryo may not be ready to "wake up." Once formed, it may go into a state of dormancy much like animal hibernation, and time as well as temperature may play a part in its awakening. There is some evidence, too, that some embryos must experience a long period of cold before they will awaken and begin to grow. This is particularly true of many temperate zone plants.

Then there is the seed coat itself. When it is new, an apple seed is hard and smooth and may be so tightly closed that it will simply not permit water — nor, for that matter, oxygen — to enter. And oxygen, of course, is necessary for the embryo to begin to respire.

Most of the above we know from one plant or another, but there are some fascinating questions that remain unanswered. Why, for example, do the seeds of some of our most important vegetable and cereal crops, as well as of some flowers, germinate only after a period of dry storage? Why do some varieties of lettuce germinate easily at cool temperatures but refuse to grow when the mercury goes above 75°F? And how does light affect germination — as it apparently does in some plants?

The factors controlling germination are still incompletely understood. We know that exposure to light, varying temperature, and moisture all play important roles in the process. It appears that some internal chemical barrier or physical mechanism is present, preventing germination until the right conditions are met. Exactly how the absence or presence of certain conditions (such as cold or water or light) can trigger germination is still a puzzle.

Cabbages

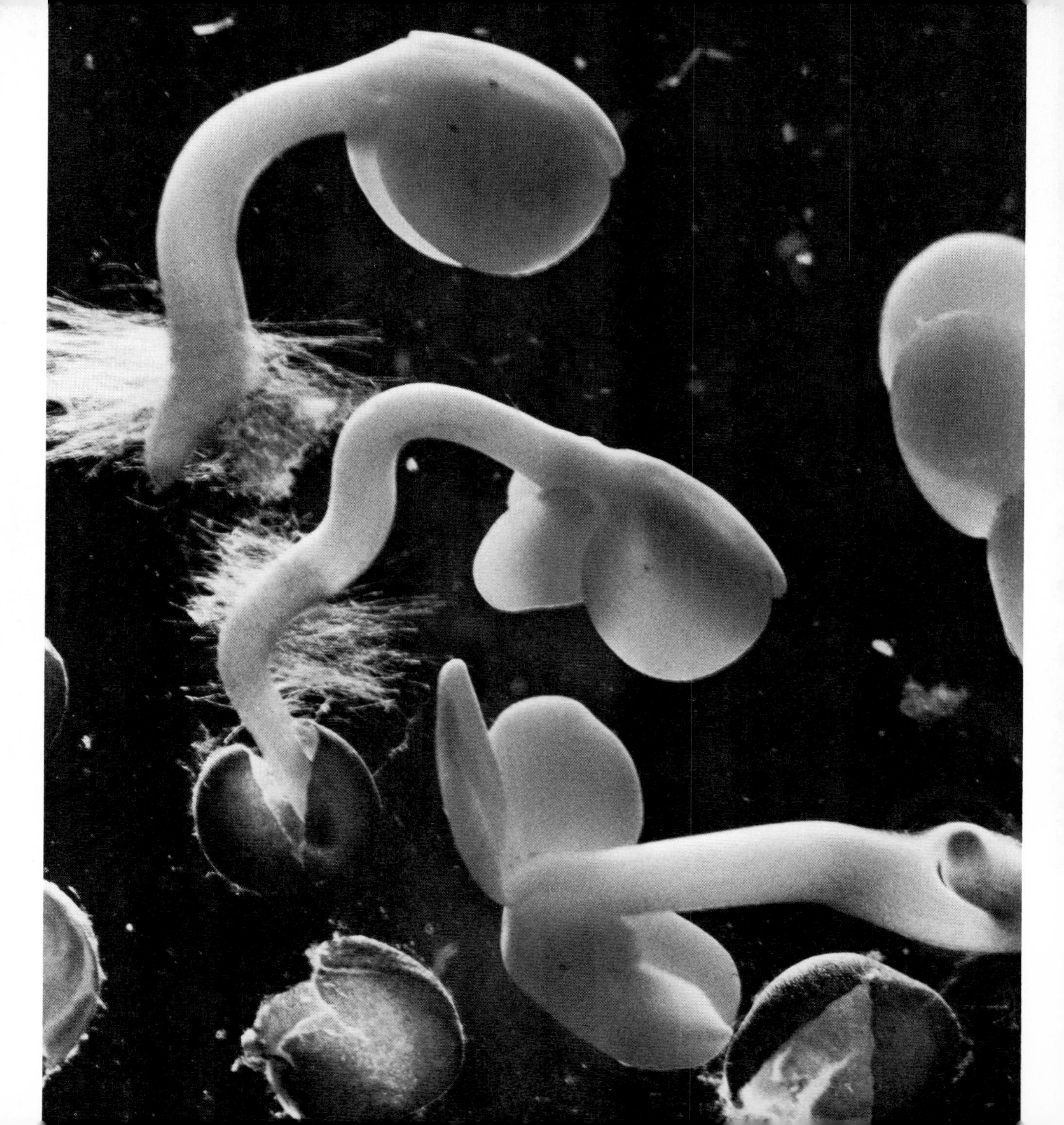

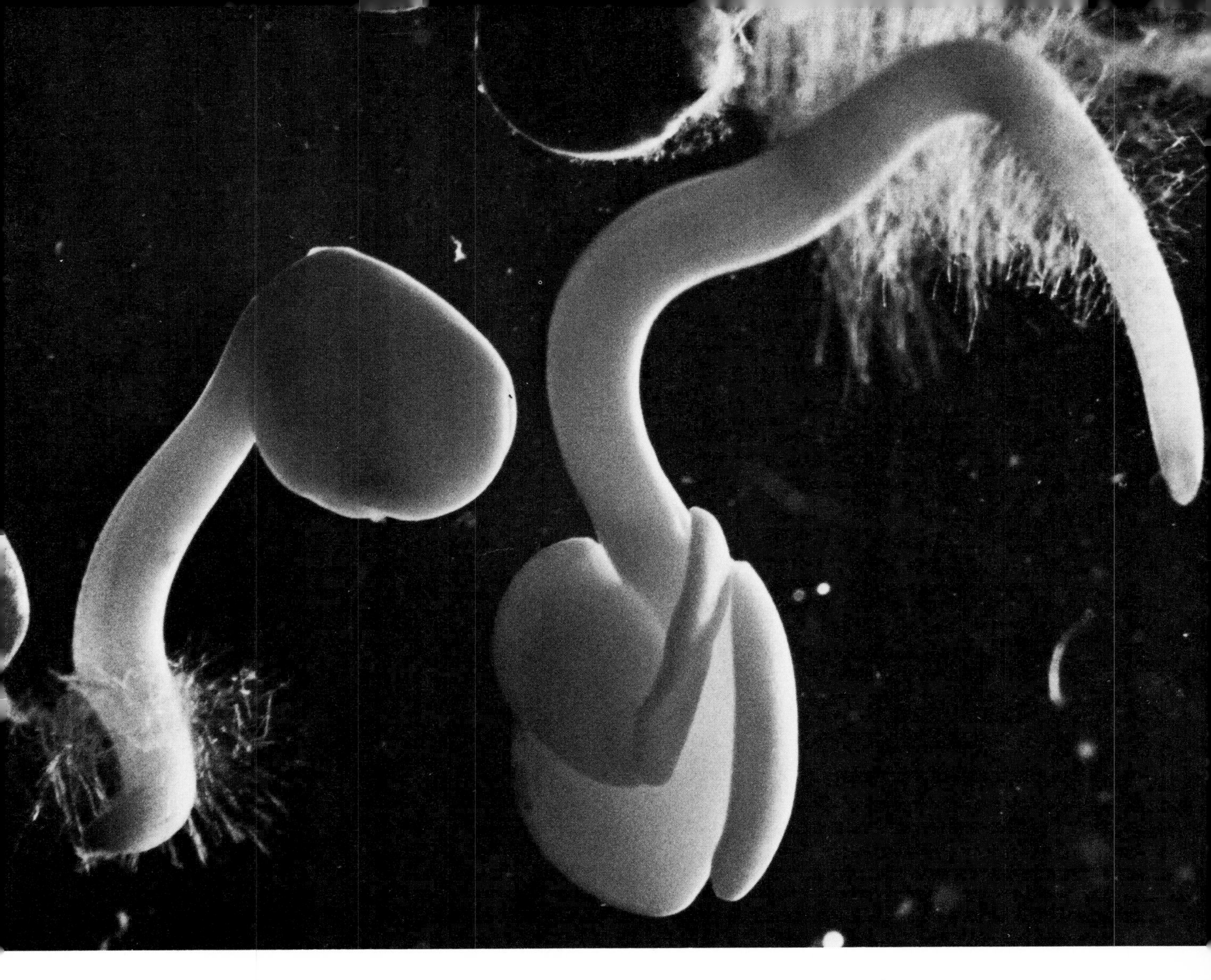

Cabbages exist in an almost infinite variety of color, shape, and size. These pale sprouts of a Chinese cabbage are still drawing nourishment from their cotyledons — the tiny, split structures from which they have sprung — but will soon develop chlorophyll cells, turn green, and begin to photosynthesize their food.

The axillary buds along a cabbage stem (below) show how Brussels sprouts develop. Brussels sprouts are a small variety of cabbage, and grow as tightly-packed buds along a stem which is topped by a cluster of photosynthesizing leaves.

Of all the cabbages, the red cabbage is the closest relative to their common ancestor, the wild sea cabbage. This red cabbage plant (right) has already put out roots; the curving growth of its stem shows how it will push through the thin soil to the open air. There it will straighten and begin to develop the close-packed, waxy leaves characteristic of all cabbages. The wax is a protection its ancestor evolved against salt spray and constant winds: it prevents water loss from the leaves.

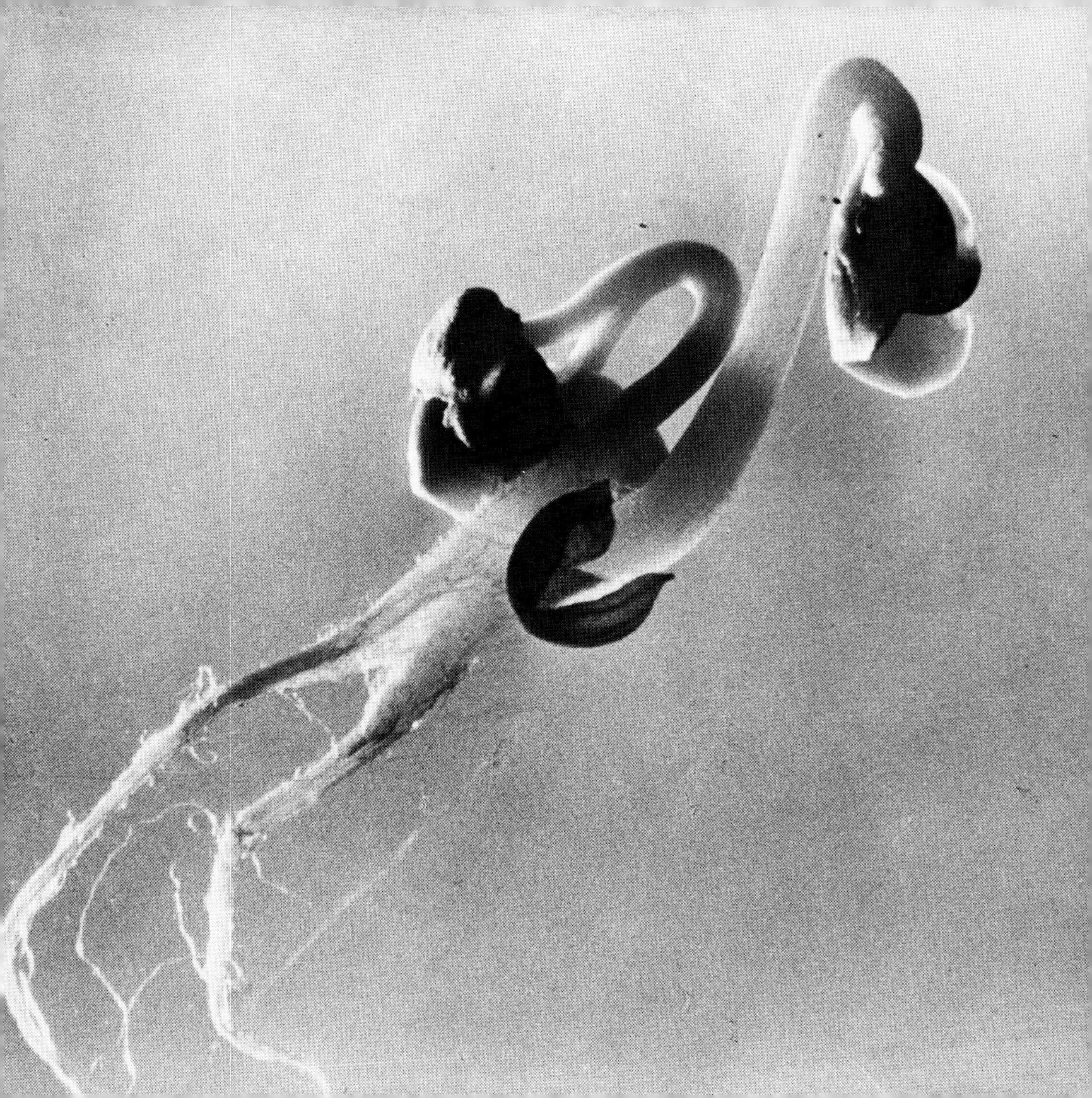

Cut through the center, this red cabbage shows how densely the leaves are packed and folded inside a cabbage head. This characteristic is inherited from the wild sea cabbage, which had to protect itself against the harshest of environments, a rocky seacoast. It is characteristic of cabbages that the outermost leaves are greenest; they do almost all the photosynthesizing. In red cabbages, anthocyanin — a red pigment in the cell — masks the green of the photosynthesizing chlorophyll.

How Cells Divide

The basis of plant growth is cell division. Through this process, a single parent cell may give rise to an almost infinite number of descendant cells. Consider, for example, how many cells there must be in a plant such as a giant Sequoia! But before we go into the details of the miraculous process of cell division, let us look more closely at what a cell itself is.

In effect, a plant cell is a tiny box, usually with four walls (sometimes there may be more, giving it a many-sided shape). In the box is a clear, viscous liquid called the *protoplasm.* The protoplasm is a highly complex liquid made up of a solution of inorganic salts, the nitrogen-containing substances called amino acids, and simple sugars, as well as *colloidal material,* a suspension of tiny particles — consisting mainly of protein molecules and fat globules — in water. Although the walls of plant cells are thin, they are also remarkably strong, being constructed of several layers of the tough, fibrous substance known as cellulose. They are also flexible, since they contain pectin, the stuff that makes jams and jellies jell. In the case of cells, pectin helps them to stick together.

How do cells divide? Simply enough — they grow a new cell wall right through the middle of an existing cell, splitting the cell in half and dividing up its contents evenly on either side of the wall. Where the process grows complicated, though, is in the division of the nucleus, the single most important item in a cell's contents because it directs the functions of the cell life and carries onward from cell to cell the unique genetic characteristics of the original parent.

The nucleus is surrounded by a *semipermeable membrane,* one that permits only certain materials to pass in and out. Within the nucleus are molecules of two types of *nucleic acids.* One of these is called deoxyribonucleic acid (DNA for short), the other, ribonucleic acid (or RNA). It is DNA which carries the genetic code — the chemical map that determines every single characteristic of a living organism — in tiny, threadlike structures called *chromosomes.* Along the chromosomes are the beadlike clusters called *genes* — each of which contains the specific instructions for a particular plant characteristic. In the course of cell division, the chromosomes divide again and again, hundreds of thousands or millions of times, and it appears miraculous that they do this so many times and yet so rarely make a mistake.

The process of cell division — or *mitosis* — begins when the chromosomes cluster together near the center of a cell. When seen under the microscope at this point, they give the appearance of a bundle of spaghetti-like filaments. The chromosomes next begin to divide, with each splitting lengthwise down the middle. The identical chromosome halves that are left by this process are called *chromatids.*

While the chromosomes are dividing, the semi-permeable wall of the nucleus containing the chromosomes has dissolved, and thin, nearly invisible fibers have started to grow, from the cell walls at opposite sides of the cell, toward the chromatids. These nearly invisible *spindle fibers* attach themselves to the chromatids, gradually pulling the chromatid bundle apart. When this process (which takes only a few minutes) is completed, two new sets of chromosomes have been pulled to opposite sides of the cell. Next, a fluid membrane, called a *cell plate,* forms between the two chromosome sets and grows gradually outward to the walls of the cell. In a matter of minutes this membrane will have divided the original cell into two daughter cells, each with its own complete set of chromosomes. Finally, the cell plate will evolve into a new cell wall. Each one of the daughter cells can then repeat the division process, reproducing itself. Thus, from one original cell there first come two new cells; from these two come four; from the four come eight, and so the multiplication goes on for as long as cell division continues.

Peas

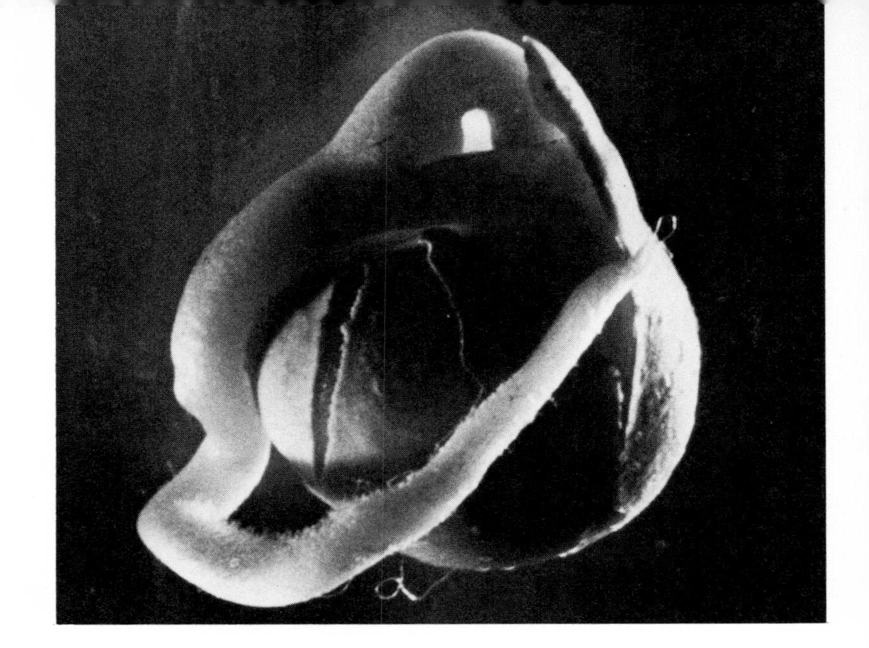

Few vegetables could be more familiar to us than garden peas: but these are peas as we would never see them unless we grew them ourselves. The pea we eat is the endosperm, or food supply, for the pea embryo. The first growing effort of the embryo is the emergence of the hypocotyl (above and left), from which the root will grow. On the opposite page, the epicotyl, which will form the shoot, grows up to seek light,
while the hypocotyl continues to develop. Here, it has already begun to put out lateral roots.

When a young pea plant is mature, its roots will cover many yards of ground and — like many legumes — will probably have developed small root nodules, gall-like structures caused by bacteria invading root hairs and infecting the root system. Fortunately, these are nitrogen-fixing bacteria which take nitrogen from the soil and convert it to a form that can be used by the plant. Plowed under at the end of the season, the plants will thus add nitrogen to the soil, reducing the need for artificial fertilization.

Here is the shoot of a simple pea plant with its cotyledons — the large, dark objects in the middle of this tangle. They protect and nourish the new shoot as it grows outward from them. Below them, the branched root shoot is already well-developed.

Newly formed leaves are still unfolding below the growing tip of the pea plant — the apical meristem, where cell division takes place. At the bottom of the picture you will see the tiny tendrils with which the pea plant holds itself upright on a trellis or fence.

Here the veins of the translucent leaves of the young pea plant are clearly visible (right) — veins which will carry the products of photosynthesis to the roots to nourish them.

A pea plant has one of the loveliest of flowers, as delicate in structure as it is in color. Like most flowers, it forms in the axil of a leaf, the point where the leaf joins the branch or stem of the plant. Once the flower is fertilized, the ovary, its reason for existing, will dominate. The pretty petals will wither and fall away; the styles, stigma, carpel, and other flower parts will dry up and disappear. Only the ovary will be left, with its burden of fertilized seeds.

Where the flower once was now hangs an ovary filled with edible peas. Until the seventeenth century, these were dried before eating: it was the court of Louis XIV of France, responsible for so many gastronomic and other innovations, that first made fresh peas popular. Of course, if the peas are not harvested, the pod will dry up, split open, and release its fertilized seeds to fall upon the ground where they will germinate in the spring.

Where Cells Divide

Once you begin to ask questions about plants, it becomes very difficult to stop. The question of how plants grow, for example, seems simple enough to answer — they grow by cell division — until you start to wonder about where cell division takes place. Do cells divide at one end of a plant? Or both ends? Or all over? And what happens when they have divided?

Let's go back to the germinating seed. Germination begins when a seed absorbs a large quantity of water — usually in the spring, when the proper conditions of day-length, light, and temperature exist. The water brings the dormant plant embryo to life, and its cells begin to divide, synthesize new protoplasm, and swell up. Finally the embryo bursts from its seed coat and the new plant begins to take shape.

The embryo of a plant is built along an axis, and thus it has two ends. One of these ends, called the *epicotyl*, forms the stem and leaves of the plant; the other end, called the *hypocotyl*, forms the root. In addition, there are the *cotyledons*, also known as seedling leaves: these are attached to the embryo and, by absorbing food from the endosperm, make it available to the embryo as needed. Once the epicotyl begins to grow, the cotyledons grow along with it, affording the new shoot a measure of protection as it pushes up through the coarse earth.

The locations in a plant embryo at which cells divide, and where new growth occurs, lie at each end of the axis; the tissue in the two locations is called the *apical meristem*. At the tip of the hypocotyl, dividing cells in the apical meristem tissue form the root tips of the young plant. These root tips form behind a *root cap* of tough, dead cells which bulldozes its way through the soil, pushed forward by the dividing cells behind it. Since the root cap takes so much punishment, it is constantly being worn away; it is constantly replaced, however, by new cells formed behind it, which grow thick walls and move to the front to form a new root cap.

At the epicotyl — the other end of the embryo plant — cell division produces a different kind of cell, the delicate, light-green cells of the shoot, containing chlorophyll. These cells will eventually produce the branches and leaves of the plant.

Protected by the seedling leaves, or cotyledons, the shoot very quickly pushes through the earth and reaches the air. Since the root must furnish nourishment for the growing plant, and hence should be a step ahead, the hypocotyl of a plant embryo usually begins growing before the epicotyl.

This is not, however, the full story of plant growth; it is only the beginning. Other aspects of the growth and life of a plant will be explained as we move along.

Potatoes

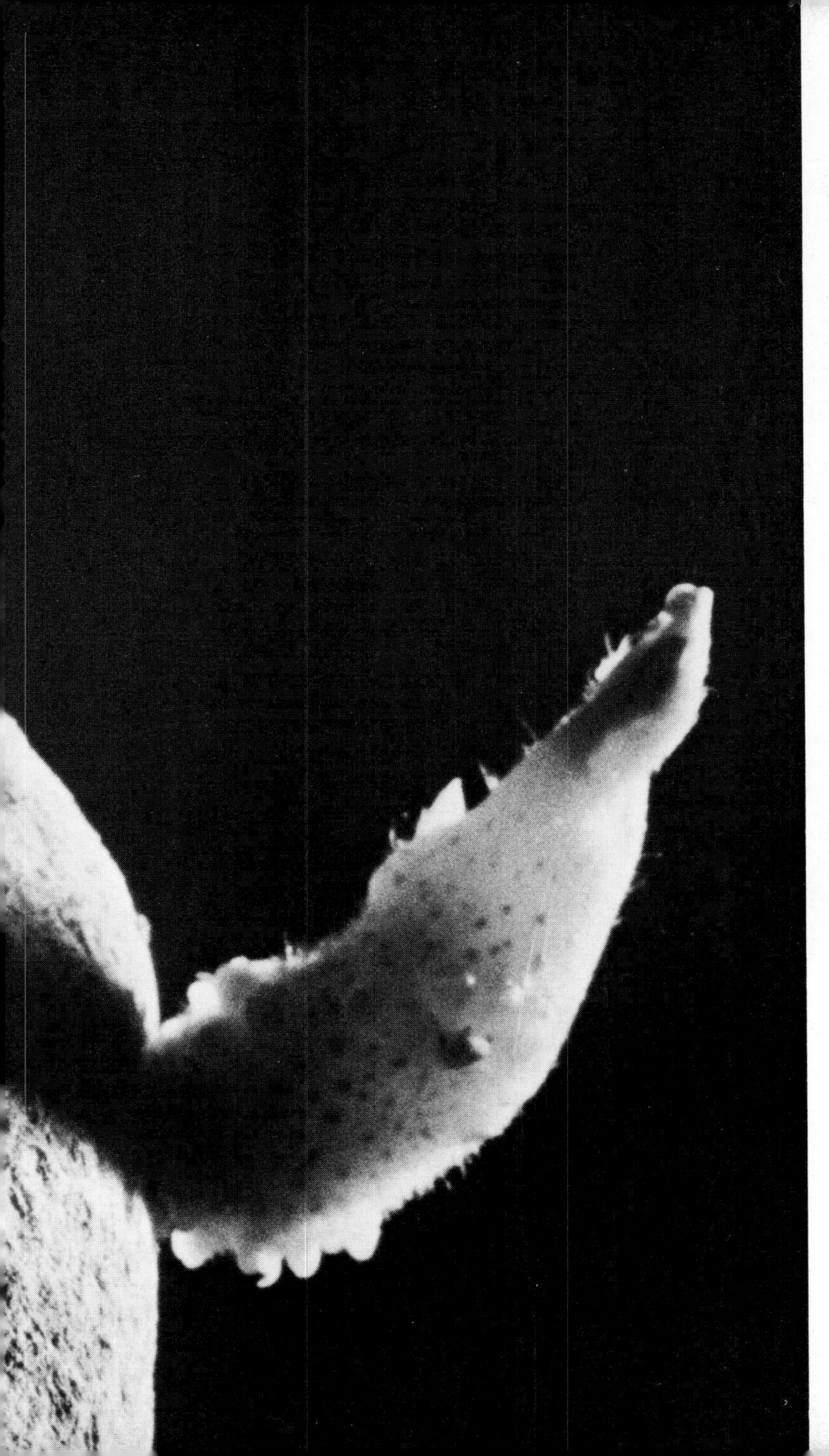

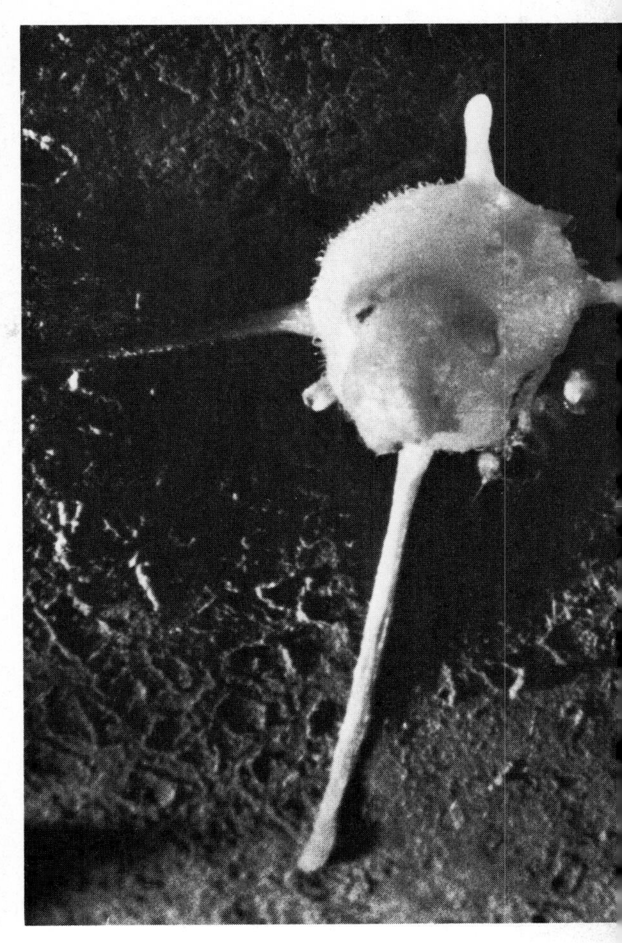

Potato plants sprout directly from the tuber (left), the part of the potato we eat, which grows underground.
The tuber is really a grossly swollen stem, used by the plant as a storehouse for food, particularly for starch. The new shoots growing from it (above) will form both stems and roots.

The two long "tails" are adventitious roots, so-called because they sprout directly from the stem of the potato plant.

This wild tangle of growth indicates the extraordinary ways a potato has of propagating itself. It can be grown from seeds, although it generally reproduces itself from buds on the piece of thickened stem which forms the potato itself.

Stolons form among the shoots arising from the potato tuber: they are the long, thin runners threading among the shoots. The stolons have buds at intervals along their length and at their tips; and it is these buds — as these pictures show — that develop into new potatoes, growing in size and thickness as they absorb food from the rest of the plant.

Like the tomato, the potato plant is a relative of the deadly nightshade — a plant known for centuries for its poisonous properties. But none other than Marie Antoinette treasured the potato for its pretty white and yellow flowers, which she wore in her hair.

Nestled among an abundance of small, shiny, dark-green leaves, the potato blossom has a delicate beauty which utterly belies the grossly swollen tuber underground. The blossom comes late, after the tubers have begun to form, and in the wild it serves to propagate the plant by its tiny seeds.

The Mechanics of Growth

Cell division, by itself, is not the only way in which a plant grows; if it were, plants would take far too long to sprout and develop. Thus, although cell division in the apical meristem does create new plant tissue, some 90 percent of the actual growth of a plant takes place through elongation of cells.

The area in which this elongation takes place lies behind the apical meristem, in what is known as the *zone of elongation*. Here the cells are still new, fresh, and pliable; and, as any plant grower knows, they absorb a great deal of water. They don't merely take up as much water as they need; they take up even more, and as a consequence, have to stretch to accommodate it all.

There are two surprising things about this stretching. One is that the cells do not greatly increase in girth. Most of their stretching is lengthwise, which means that they primarily grow longer. The second surprising thing about the stretching of cells is that, having grown as long as it can, a cell cannot get smaller again: its growth is irreversible.

This seems logical enough: we hardly expect to see plants get bigger or smaller according to their available water supply. But for a long time nobody knew why this was so, because it was impossible to study the construction of the tiny cell walls. Not until the invention of the electron microscope was it possible to examine how a cell wall is built, and then the questions of one-way growth and irreversible growth were answered very clearly.

Like everything else in nature, the structure of a cell wall is beautifully functional, and the wall goes through various phases to meet changing needs. Remember that in a very young cell, one that has only just divided, the new cell wall begins with a cell plate which grows outward to the old cell walls, dividing the cell into two equal parts. Once this division has been accomplished, the new cell wall, formed from the cell plate, congeals into a thin, randomly-woven network of threadlike structures called fibrils. This is the *primary cell wall,* and it is the foundation, so to speak, for the secondary cell wall which permanently divides the two daughter cells.

The secondary wall is not randomly woven; its fibrils are laid down in neat horizontal rows around the girth of the cell. Furthermore, they are made of cellulose, a tough, rigid material which has no give to it. The cellulose rows, however, are embedded in pectin, and pectin can stretch. But because the fibrils run around the girth of the cell, the walls can stretch in only one direction — lengthwise. Finally, hardening takes place when a substance called lignin appears in the structure of the expanded cell walls, making them rigid. And this is why plants grow along an axis — because millions upon millions of their cells in the zone of elongation are stretching, pushing outward to the limits of their capacity, then hardening so that this growth is permanent.

Tomatoes

Here is a mass of tomato seeds, each of which contains everything required for the development of a complete tomato plant. When the ovary — which is the tomato we eat — has ripened and fallen to the earth, the seeds will be mature, and if conditions are right, the process of growth will begin.

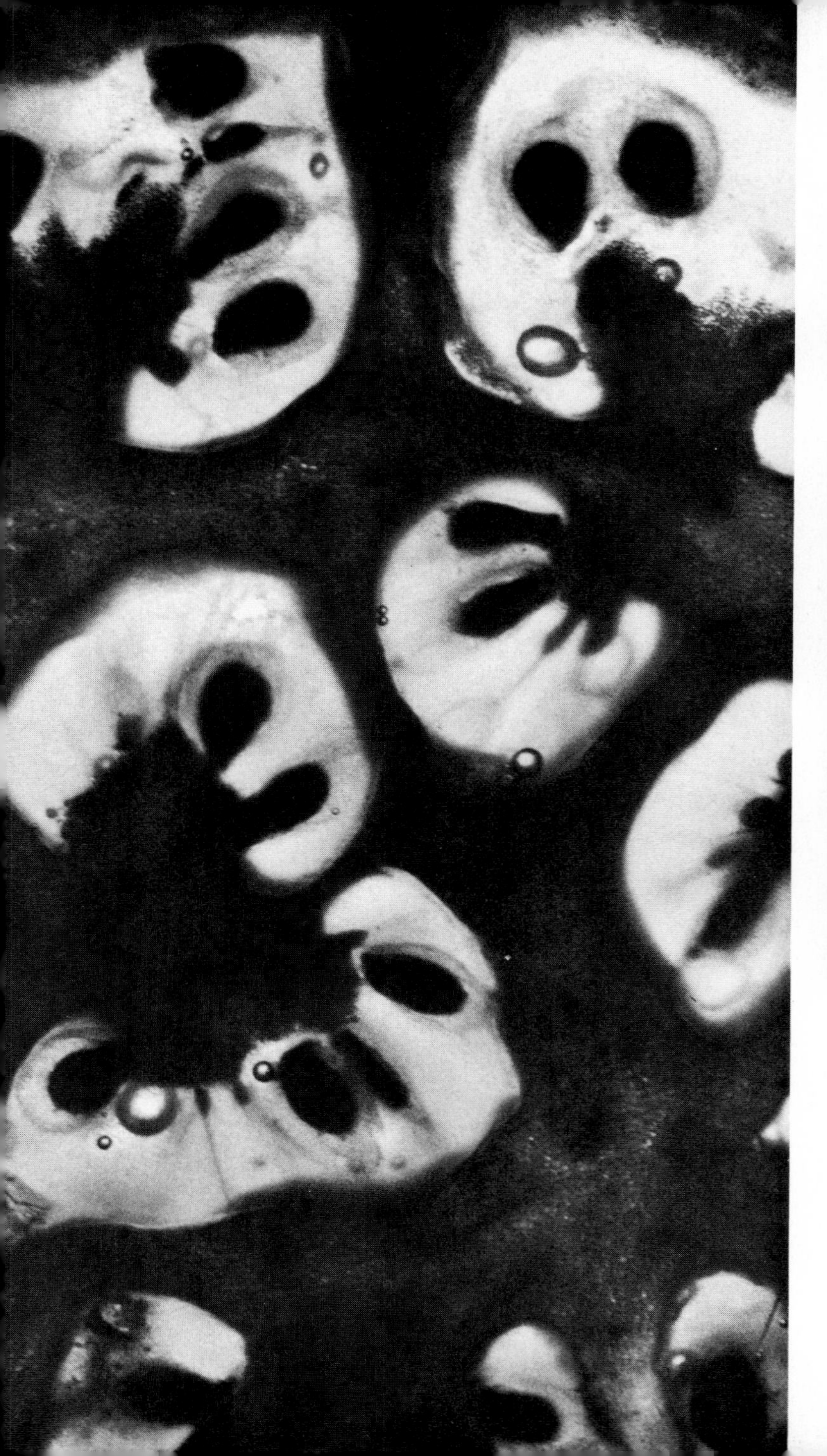

These slices of tomato reveal the seeds suspended in the protective ovary.

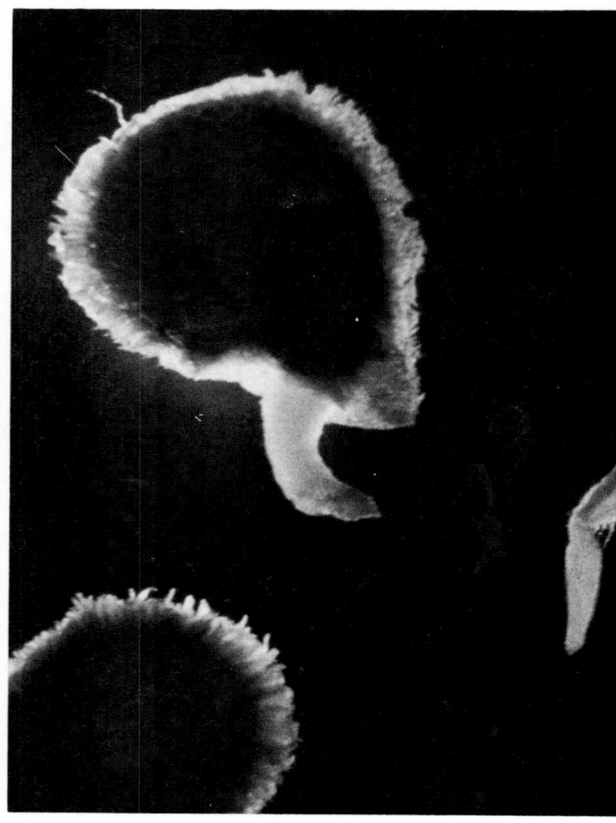

These tiny creatures are already prepared to start life on their own. The little tails they have developed are their hypocotyls or newly formed roots: these will probe deep into the earth and supply the growing plant with all the nourishment it needs until it has put forth leaves, which can then take energy directly from the sun and convert it to plant sugars.

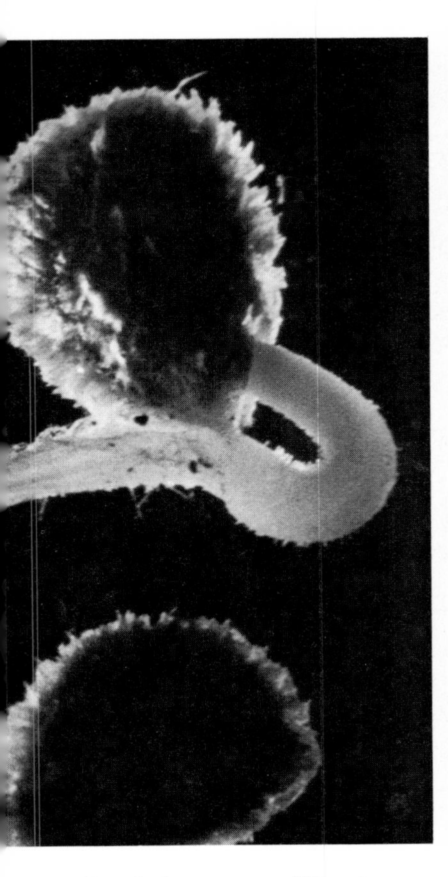

Cotyledons, or seedling leaves, nourish and protect the embryo during early stages of its development. The first green shoots have appeared within the cotyledons' protecting arms. It is from the epicotyl, or shoots of the embryo, above the cotyledons, hat the tomato plant itself will grow.

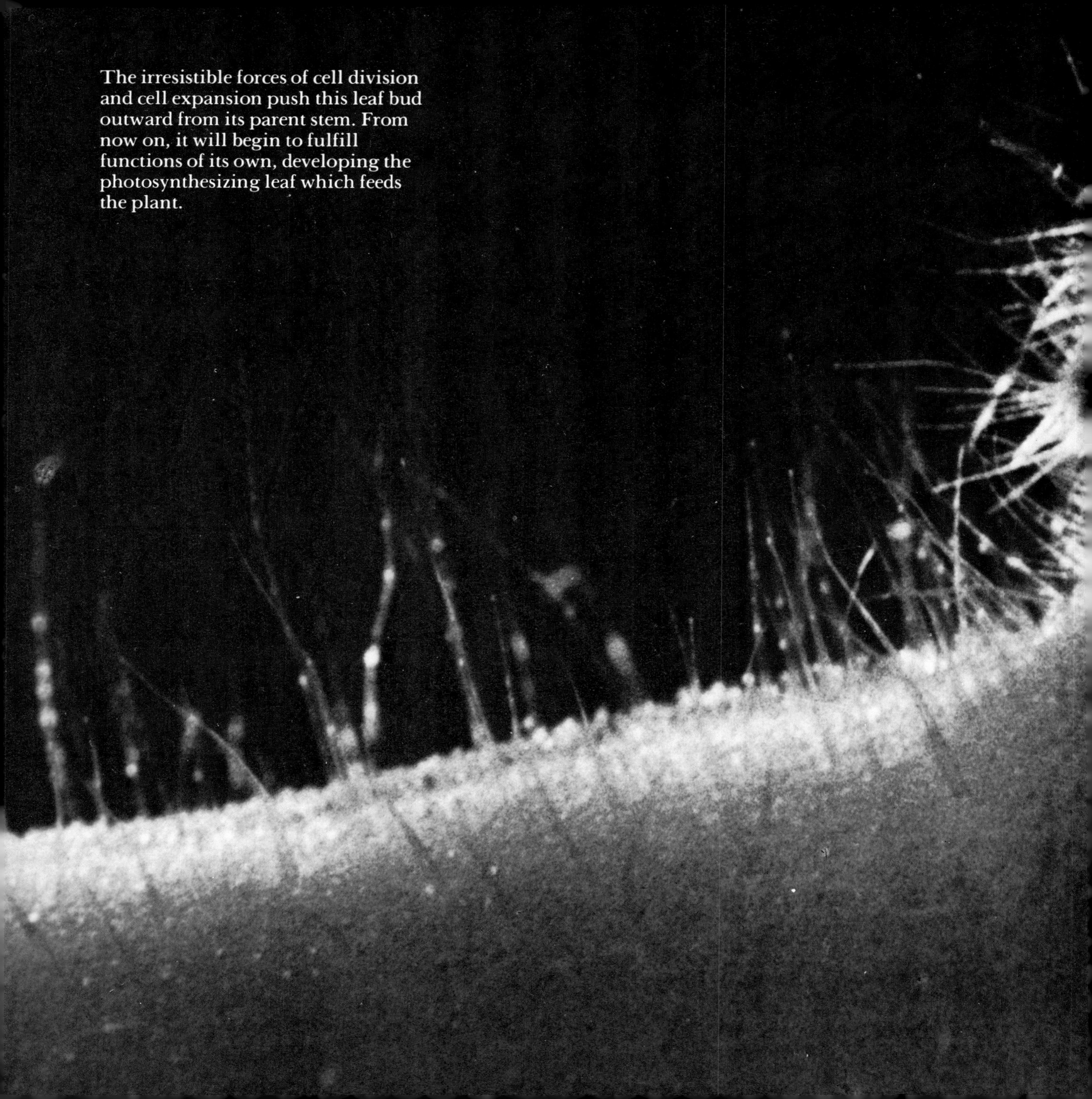

The irresistible forces of cell division and cell expansion push this leaf bud outward from its parent stem. From now on, it will begin to fulfill functions of its own, developing the photosynthesizing leaf which feeds the plant.

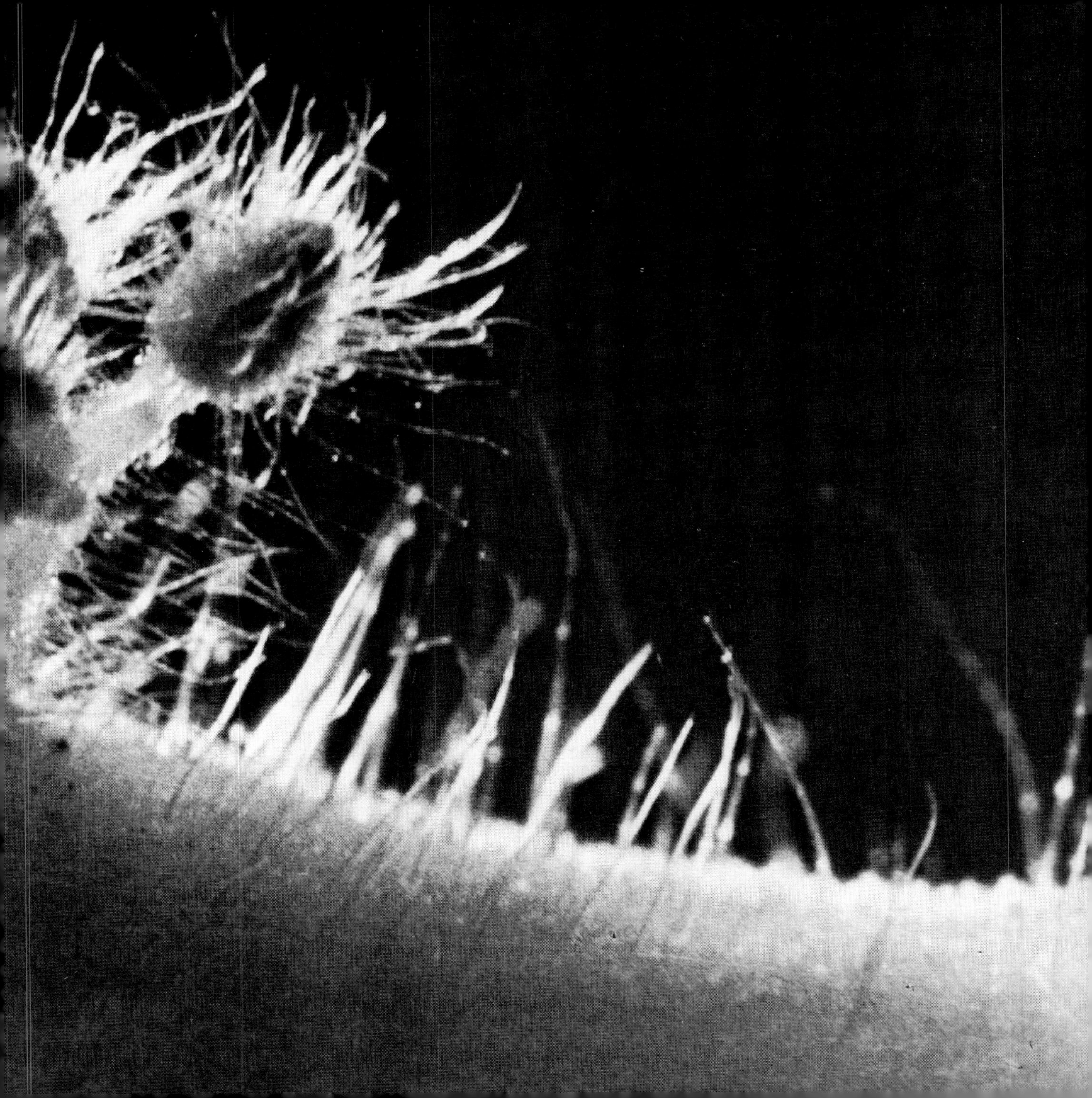

The trunks and branches of the tomato plant must be sturdy—as these are—for its fruit is large, pulpy, and heavy enough to bend even the strongest branch.

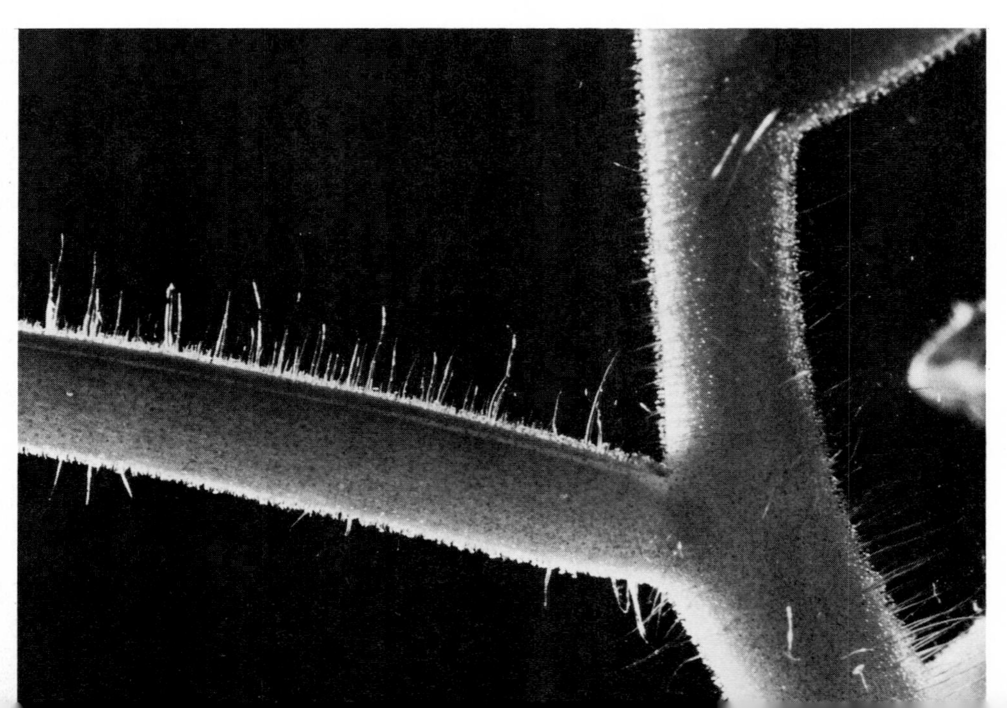

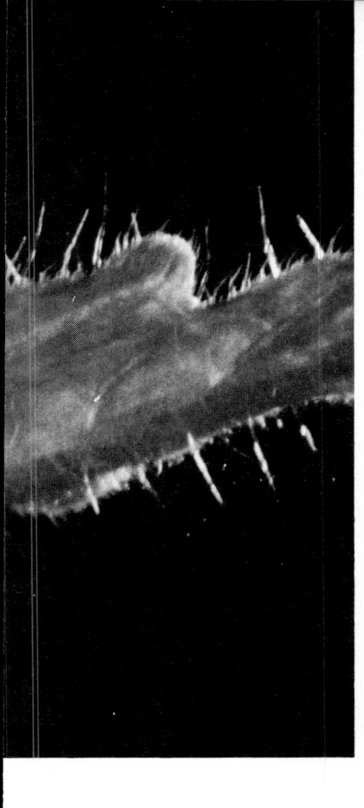

Why all these hairs on the stem, branches, leaves, and flowers of the tomato plant? They protect against excessive water loss in the hot South American climate where the tomato originated. Water droplets trapped by the hairs as the plant transpires help it to keep cool.

If you look at a tomato in the supermarket, you will see at one end the remains of its attachment to the plant — sometimes even a piece of stem and a bit of leaf — and at the other end a small, dimpled scar. That tiny scar is all that remains of the style of the tomato blossom from which the fruit came.

Phototropism

Plants, you may think, have problems: How do they know in which direction to grow? How can they tell up from down? How can they tell light from dark? How does a growing plant know when to put out a branch, or a leaf?

Obviously, they do know, for if they didn't, the plant world would be a scene of wild confusion. As it is, when a seed begins to sprout, the root (which usually appears first) immediately turns downward and starts growing into the soil; the shoot, when it appears, does just the opposite, growing upward toward the light. Moreover, once the growth has started, it follows very strict patterns for each species of plant involved, so strict that we can invariably tell one species of plant from another by the shape of its leaves, the arrangement of its branches, and other such outward signs.

The eminent discoverer of adaptive evolution, Charles Darwin, was one of the first scientists to address the fascinating question of how plants grow, about 100 years ago. He wanted to know what made plants bend toward the light and that they do, anyone can see simply by turning a houseplant away from the light and observing how it then begins bending back toward the light. In his simple but elegant experiments, Darwin decided to use the *coleoptiles* of grass plants. Coleoptiles are tubelike cylinders that grass plants put forth to protect the first tender seedling leaves as they push their way up through the abrasive soil. The coleoptiles grow very fast, and also show a strong tendency to turn toward the light, a phenomenon called *phototropism,* from the Greek words for light and turning.

For his experiment, Darwin cut the tip off a coleoptile; the coleoptile stopped bending toward the light and also stopped growing. Next, he covered the tip of another coleoptile with a lightproof cap: the result was the same. But when he covered the tip of still another coleoptile with a transparent cap, it kept on bending toward the light and also kept on growing. Finally, just to make certain that the seat of the action was really in the tip, where it certainly seemed to be, Darwin slipped a piece of lightproof tubing over the bottom part of the coleoptile, where the actual bending occurred; this did not stop the bending, nor, indeed, did it affect the growth of the coleoptile at all.

From these experiments Darwin drew the conclusion that the tip of the coleoptile was sensitive to light, and that "some influence is transmitted from the upper to the lower part, causing the latter to bend." What that "influence" might be, Darwin did not venture to guess, but he had laid the foundation for further experimentation.

Thirty years after Darwin's experiment, another scientist came a step closer to defining the "influence" of which Darwin had spoken. P. Boysen-Jensen wanted to see whether the "influence" was chemical, electrical, or nervous in origin. He took some oat coleoptiles, cut off their tips, then placed a thin piece of gelatin on top of the cut. As might have been expected, the coleoptiles stopped growing; but Boysen-Jensen's next step was to put the amputated tips back on the coleoptiles — on top of the gelatin. There now existed a neutral — but permeable — barrier between the coleoptile and its amputated tip. What happened?

The coleoptiles resumed growing as though nothing had changed. Not only that, but they bent toward a light source shining at them from one side. In other words, the tip had received the light message, and this message had somehow moved through the gelatin strip and down to the base of the coleoptile, where the bending had been induced. That this message could have been transmitted electrically or nervously was still not impossible, but it seemed highly unlikely: a chemical messenger was a much more logical agent.

In 1918, a Hungarian botanist stepped into the picture to demonstrate that a coleoptile tip could influence a plant's bending even in total darkness. A. Paal cut off the tip of a coleoptile, then put it back on again — but off-center. He found that putting the tip back on the left half of the coleoptile caused the plant to bend to the right; putting it back on the right half made the plant bend to the left. This indicated that the part of the coleoptile directly under the replaced tip was stimulated to grow fastest — another strong indicator that a chemical agent was at work.

Eight years later a Dutch scientist named Frits Went finally proved the chemical hypothesis once and for all. Went cut the tips off coleoptiles and put them, base down, on small blocks of agar, letting them remain in this position for about an hour. Agar, made from seaweed, is a gelatin-like material with properties of absorption and purity that make it an ideal laboratory culture medium. Went then removed the coleoptile tips from the agar and put the agar blocks, minus the tips, back on several coleoptiles. The results were in every way the same as if Went had replaced the coleoptile tips themselves: the coleoptiles began to grow again. They bent toward the light, and if the agar blocks were placed off-center, as in Paal's experiment, the coleoptiles bent even in the dark.

Obviously, a growth-inducing agent had diffused from the tips of the coleoptiles into the agar, and equally obviously, this growth-inducing agent must be a chemical. Furthermore, since the agent was a chemical that had been produced in one location (the coleoptile tip) and had acted in another location (the coleoptile base), it qualified as a *hormone* (or plant growth substance), and Went named it *auxin*, from a Greek word meaning "to grow."

In recent years, the mechanism of how auxin causes plants to bend toward light has been explored. Auxin itself is affected negatively by light—it is fairly certain that light destroys auxin, thus leaving active auxin only on the shaded side of a plant, with the result that this side of the plant grows faster, causing the plant to bend toward the light.

Auxin is also involved in the directional growth of roots and shoots. In this case, the stimulus involved is not light but gravity, and the phenomenon is called *geotropism*. Moreover, there is a very subtle difference in the effect of auxin in shoots and roots, which demonstrates a degree of sophistication that is truly astonishing.

When a seed sprouts and sends out a shoot, auxin is produced in the shoot tip, and it appears that gravity causes it to accumulate on the bottom side of the shoot. This stimulates growth on this side, with the result that the shoot turns upward and starts growing toward the light. But what about the root?

Here, too, auxin is produced, in the tip of the root called the root cap; and here auxin also accumulates on the bottom. Yet the root does not turn upward, as does the shoot; it turns down! Why?

This took some rather deep and complex investigation, and a general answer has been found. Auxin does accumulate on the bottom side of the root; it has been measured. But the action of auxin in the root is governed by two opposite effects of this plant hormone. In too great a concentration, auxin inhibits growth of the root while in small concentrations it stimulates growth. Thus, in the upper part of the root, where the least amount of auxin is present, there is the greatest growth, and consequently a bending downward of the root. The least amount of growth occurs in the bottom of the root, where the most auxin is found. In the course of a plant's life, its roots twist and turn many times, sometimes growing upward, sometimes downward, and the highly sophisticated reaction of the root to auxin may well be the answer to these erratic ramblings of the root.

African Violets

This photograph illustrates one of the African violet's most beguiling qualities — its self-renewal from a leaf cutting. When you plant a leaf (or even just part of the leaf) of an African violet, adventitious roots will form below the ground, and you will soon have a new little plant growing against the backdrop of its parent, as here.

The African violet has a bisexual flower, containing both the male and female organs of reproduction, and in these remarkable photographs both sexes are shown. In a cluster of four round sacs are the pollen-producing anthers: pollinating insects will rub against these while seeking nectar in the flower and carry away a golden dusting of pollen to the next plant they visit. The stalk or style, pointing upward next to the anthers, is not part of the male sex organs, but of the female: at the tip of this style is the sticky stigma which catches pollen from the nectar-seeking insect or the wind.

Another view of an African violet is shown as it grows upward out of the hair-covered corolla which encloses the ovary. Remember that when a pollen grain has been deposited on the stigma at the tip of the style, it grows a tube which penetrates the full length of the style into the ovary, where a sperm nucleus from the pollen grain fertilizes the egg.

On some trees, a bud must protect the flower throughout an entire winter, and the buds of such trees have scales — which are really specialized leaves — to do this job. In the African violet the situation is less extreme, but the principal protective device of the buds can nonetheless be clearly seen here: myriad tiny hairs protect the tender buds from drying winds, and make them unappetizing to a predator.

This African violet plant has grown and put out buds in preparation for flowering. Buds are primarily protective structures for the flower, that all-important reproductive organ that develops inside them.

As the blossom begins to unfold, its structure can be seen clearly. On the outside are the modified leaf bases which, during budding, tightly enclose the whole flower and all of its delicate structures. Next are the petals, also modified leaves, colored in order to attract insects. The style, with the stigma at its tip, is dimly visible in the center of the flower. Soon the flower will be fully open and another life cycle can begin.

Cell Differentiation

The embryo of a plant is a miraculous bit of tissue. Within its tiny self it carries the potentialities for an entire plant, complete with leaves, branches, blossoms, and roots. And depending on the species to which it belongs, the plant can be anywhere from an inch or two to 350 feet tall.

All of the different parts of a plant involve different kinds of cells, some of them extremely complex, such as those in the leaves which carry out photosynthesis (page 117). Where and when do cells such as these develop in a plant?

The cells that do the various jobs in a plant are formed by a process known as differentiation. Differentiation, in turn, is part of a larger phenomenon called *morphogenesis,* which refers to the development of a plant's form.

Initially, differentiation takes place in the area immediately behind the zone of elongation in the developing root and stem. Here, for example, certain cells differentiate into the sieve cells which are part of the plant's plumbing system, while other cells differentiate into vessel elements, a different part of the same system. At the other end of the *radicle* — or root-tip — the outer and inner layers of the plant's root will be forming. Some of the tissues in these layers are highly specialized to grow root hairs — long, single-celled extensions of the outer root wall, which probe the soil for water and minerals.

In the shoot, the pattern of differentiation is somewhat different and more complicated. Leaves, flowers, and branches form at the tip, around the apical meristem in which cells are constantly dividing. Epidermal cells, which will make up the outer skin (or epidermis) of the plant, form in the area of differentiation, behind the zone of elongation; here, too, is formed the cortex, the tissue which lies between the outer skin and the water-carrying vessels of the inner skin of the plant.

As the stem of the plant grows longer, other areas of cell division are created at regular intervals along its length. These develop into the so-called *bud primordia,* which give rise to branches, leaves, and flowers, all of which will take shape from apical meristem tissue of their own.

Onions

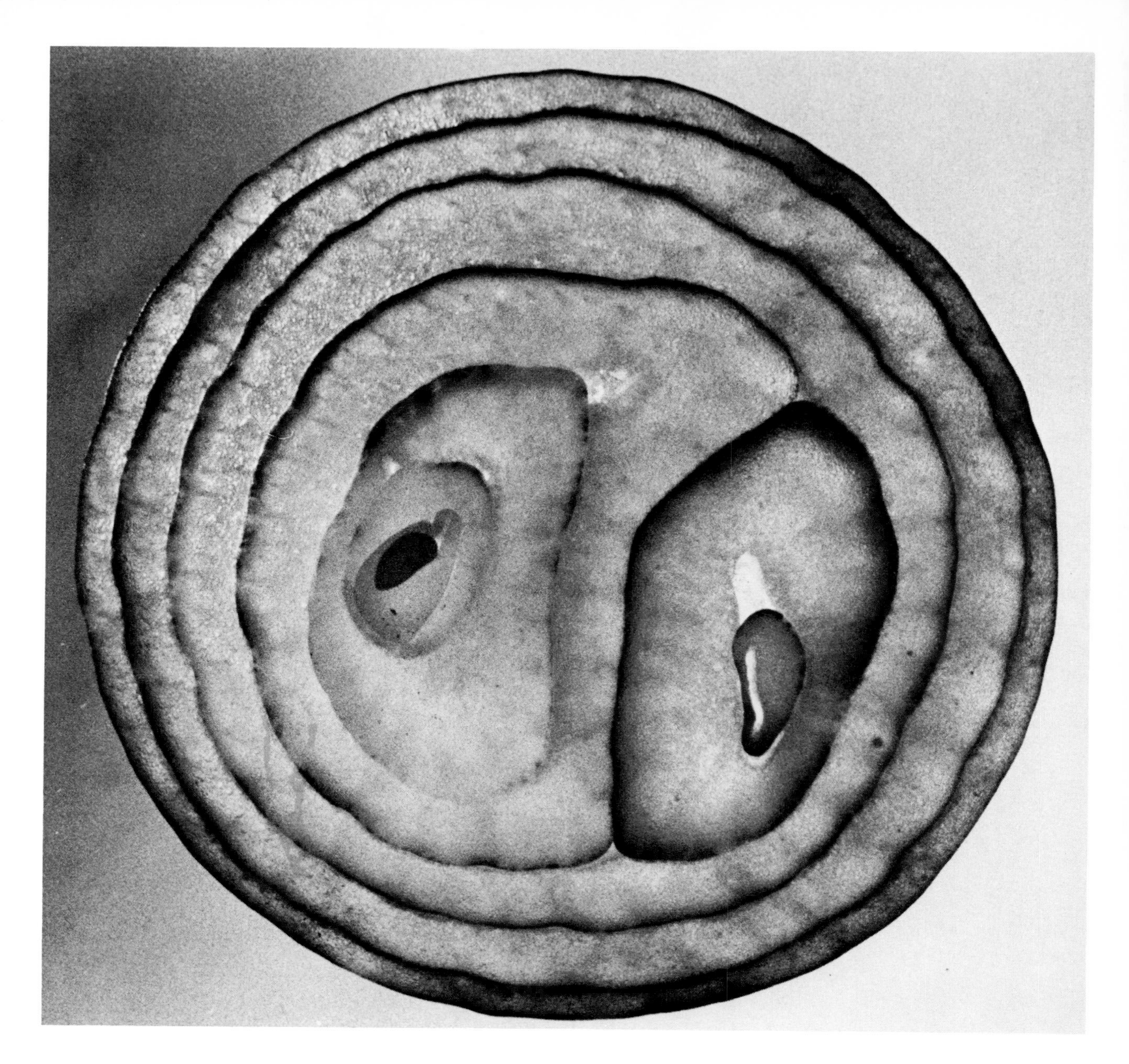

Cut an onion crosswise, and you get a model of the universe — or so thought the Egyptians, who accorded at least one variety of onion the honor of divinity. This is one of the oldest of all cultivated plants, a member of the lily family which now exists in at least ninety-four varieties.

Because it uses an underground bulb as a storehouse, it might be thought that the onion has some kinship with the potato, but this is not so. One basic difference is that, while the potato stores starch, the onion stores sugar.

At right is one type of flower produced by an onion: a bulb at the end of a seed stalk, unopened and pregnant with bulblets. This is a wild onion, growing from the bulb shown below. On the opposite page, bulblets of the same onion are visible and sprouting, graphically illustrating the kind of energy contained in this sugar-storing plant.

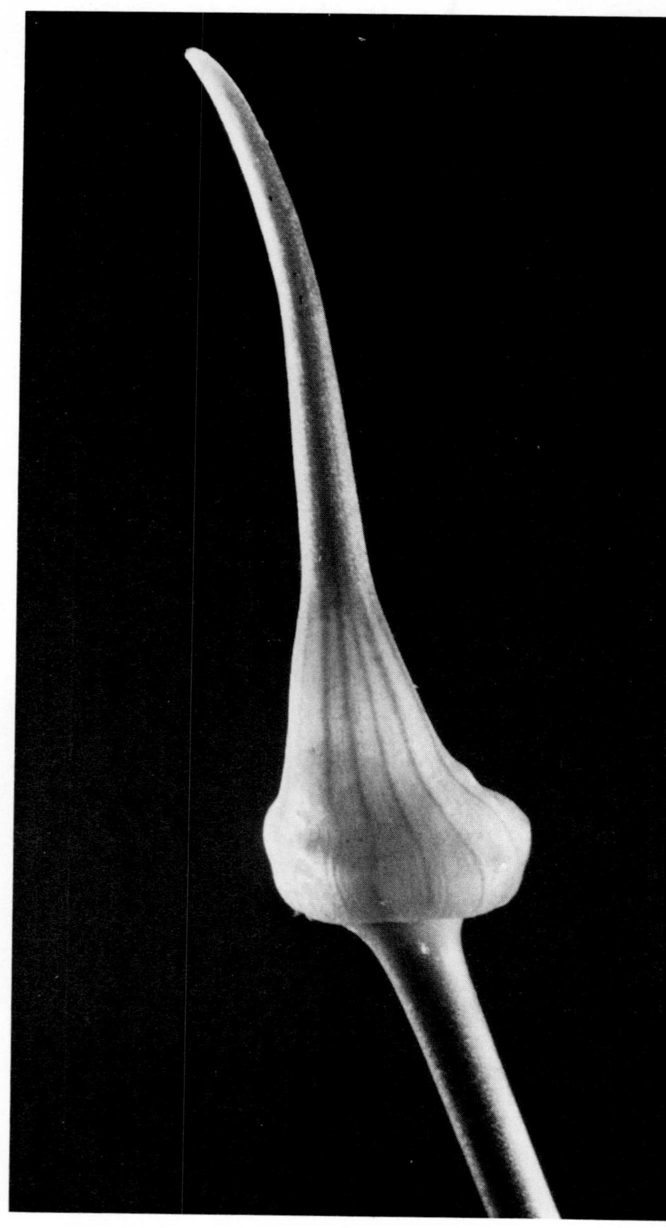

One seed stalk leads to another: here a stalk growing from a cluster of wild onion bulblets is in the process of producing a flower. This flower will blossom and produce a host of tiny, black seeds, each of which in turn can produce an onion plant. In the domesticated varieties of onion the bulblets, rather than the seeds, would be planted to produce more onions.

Here is how the onion propagates itself from a bulb. Like other bulbous plants, onions put out roots and shoots (left), fueling their growth with the sugar stored in the bulb. But when the leaves and flowers wither in the fall, there is still enough food in the bulb above to nourish the formation of a new plant, which will then appear in the spring.

Almost every part of the onion is hollow: the bulb, which is a thickened stem, is hollow at the center; the leaf stalk is hollow; the stem is hollow too. The flower cluster which forms at the top is pinkish or white and looks like a sunburst of energy radiating upward from the bulb buried in the ground.

Forming Leaves

At some point in its development, a young, growing stem must begin to assume the attributes of a plant — rather than continuing to add more stem, it must start putting out leaves and flowers. How does this happen, and how does the plant know when to do it?

We do not yet have all of the answers to the multitude of questions about plant growth; but we do have a good many, and more are being provided all the time. We know, for example, that the basic instructions determining what kind of a plant will grow from a given seed are contained in that plant's genes. We also know that the plant growth substances known as hormones are chiefly responsible for a plant's growth, particularly the hormones known as auxins. We also know that the plant hormones do not act exclusively as growth stimulators. They can, on occasion, inhibit plant growth as well. With all of this in mind, let us see how and when a plant starts putting out leaves.

How are the leaves of a plant produced? The answer is that leaves develop through cell division at the tip of the plant, where normal growth of the stem is constantly taking place. What happens is that at regular intervals, swelling forms at the base of the apical meristem, the area of active cell division at the tip of the shoot. This swelling, produced by dividing cells, forms what is called a *leaf primordium*, or bud. Like the apical meristem, the bud is an area of active cell division; but it has the capacity of making a very complex structure, quite different from the stem. This structure is a leaf.

The place where a leaf branches off from the stem is called a node, and the length of stem between successive nodes is called an internode. In nearly all plants, the nodes are very regularly spaced — so much so that the node spacing can be used to identify many plants.

Leaves form from the bud of a plant by cell division within the bud, and, when the time is ripe, by cell expansion, which pushes off the protective outer scales of the bud, permitting the young leaves inside to unfold to the sun. Once the leaves are mature and fully formed, the bud loses its capacity for cell division. At about the same time, however, another swelling containing dividing cells grows up in the angle between the leaf and the stem above it. Eventually, this swelling develops into another lateral bud, called an *axillary bud,* and it is this axillary bud which usually produces the branches of the plant.

Another word should be said here about auxins, the plant hormones that stimulate growth in the stem. It is possible that auxins play some part in determining which plants are tall and "leggy," with a lot of space between branches, and which are compact and bushy. It also appears that, in some species at least, certain auxins inhibit the growth of lateral buds, and therefore of leaves and branches. Horticulturists have learned to turn this to their advantage: by pinching off the tip of a growing stem, they cut off the auxin supply of a vigorously growing plant, with the result that the lateral buds flourish while the stem itself stops growing. Thus, instead of being tall and "leggy," the plant has the bushy, full-bodied appearance which many plant owners admire.

Squash

The tender radicle emerges from between the cotyledons of a squash seed. Already prepared to probe the soil, this frail-looking growth will form a matted and fibrous root system as the plant matures.

Here the cotyledons appear to have a hatlike shape, symbolizing their protective function. Look closely under the hat, between the folds of the shoot, and you will see tiny leaves.

A brand new leaf grows from the apical meristem of a squash seedling (below), which still wears the remnants of its cotyledons. The rapid growth of the squash is demonstrated by this leaf (right), which has already pushed up much higher than the parent stem.

A fruit is in the making here, just below these folded petals, which glow like the flame of a candle at the end of a thickened piece of stem. A squash, like a cucumber or tomato, is simply an enlarged ovary with seeds inside.

It's hard to recognize this squash as a vegetable, but it will be one soon. The flamelike top will burst into a flower with papery petals: the ovary below will thicken even more than here as it fills with nourishing pulp. The hairs are characteristic of plants that grow in bright tropical sunlight; they help keep the plant cool.

This is the male part of the squash plant which produces pollen. It is known as the anther, and a collection of anthers is called an androecium. Squash, which produce many flowers, are insect-pollinated and, as can be seen here, there is plenty of pollen available.

Amidst a tangle of curving stems and shoots, a papery flower rises majestically skyward. As evidenced by its cuplike base and the thin shoot from which it grows, this is the male part of the plant; the female parts have thickened bases which are actually the ovaries. Such a female part can be seen just left of the center of this picture, opposite a male part that is still just a bud.

The male flower is short-lived; it lasts a day or less. Then, having fulfilled its function of supplying pollen for the female part of a plant, the male flower starts to wilt and is soon gone.

In squash, as in cucumbers, the ovary appears below the flower. Once the flower is fertilized the ovary will enlarge to form a fruit which, when it is fully grown, will be harvested for the table.

Photosynthesis

When you look at the leaf of a plant, you see a green surface, usually with a central rib and with veins radiating from this rib to the perimeter of the leaf; and if you look very closely, you may perhaps see some very fine hairs and even what look like tiny pores. Although this might not appear to be a very complex structure, if you could look inside the leaf and right down to the molecular level — as scientists now can with an electron microscope — you would see the most marvelous and most productive chemical factory on earth.

What goes on inside a leaf? It is a chemical process called *photosynthesis* — the process that transforms the light energy of the sun into the chemical energy of plant substances. As we have already learned, this chemical reaction is one of the foundations of all life on earth. With it, plants capture energy and store it; animals then feed on the plants, and so take into themselves the chemical energy of the plant substances, which they then pass on to other animals, and so forth.

But what is photosynthesis? And how does it work? In greatly simplified terms, photosynthesis is a process that combines the carbon from carbon dioxide in the air with hydrogen, taken from water, to make sugar compounds that can be used by plants and animals as food. Photosynthesis is therefore, in a sense, the opposite of respiration: it takes carbon dioxide from the air and returns oxygen to it. This is why plants are largely responsible for our being able to breathe.

This much has been known about photosynthesis for more than 100 years. In fact, when photosynthesis was first discovered, it became common practice to fill sickrooms with flowering plants to purify the air. Today we know a good deal more about this remarkable process — for example, that it is the green parts of a plant which carry out photosynthesis. This means that some photosynthesis is even carried out in the stem, although it is the leaves which are most highly organized for this purpose. We even know the chemical formulae for the substances produced in photosynthesis and the chemical reactions that take place. And through the electron microscope, researchers have identified the structures in the leaf which carry out photosynthesis: the chloroplasts, bearers of the green substance called chlorophyll. They are shaped like footballs, and they are able to position themselves inside their cells so as to catch the most light.

The products the chloroplasts manufacture — principally energy-rich glucose — are carried through the plant and back to the roots in a transport system called the *phloem,* which lies near the outside of the stem structure. The phloem is composed of cells, placed roughly end-to-end, which form fine tubes running the length of the stem and function almost like pipes, carrying the products manufactured by photosynthesis.

One of the most surprising things about photosynthesis concerns which plants are most actively engaged in it. We might tend to think that the green land plants, which we see everywhere we look, are the most important photosynthesizers. Not so: they carry out only about 10 percent of all the photosynthesis conducted on earth. The remaining 90 percent takes place where we never even see it — in the ocean, carried out by microscopic algae which do not even have true leaves, yet are one of the most basic elements in the food chain.

Cornflowers

Two cornflower seeds take on the appearance of deep-sea creatures swimming in a burst of bubbles. Actually neither seed is larger than the head of a pin.

Air bubbles trapped in the hairs of these cornflower seeds glitter in their watery environment. The feathery fringe is an adaptation which gives the seed mobility, permitting the wind to pick it up and blow it to receptive soil.

The seed coat has ruptured and the birth of a cornflower takes place. The radicle (above) emerges and begins to grow, forming an arch like a buttonhook (right). From within the seed (opposite page) come the cotyledons, the seedling leaves which at this stage give the baby plant the nourishment it needs for growth.

The flower of the cornflower conceals a marvelously intricate arrangement for reproduction. It is actually an inflorescence, that is, a collection of flowerets and other reproductive organs. The blue flowerets in this case are neuter, having as their sole function the attraction of insects. In the depths of the flowering structure are nectaries which produce the sweet syrup the insects seek. As insects (in the case of cornflowers, usually bees) probe for this nectar, their underbodies brush against the pollen-laden anthers, which in this picture show up as bright clumps of powdery substance on the end of the dark-colored stalks, or filaments. Since cornflowers are also wind-pollinated, there is an over-abundance of pollen.

A more beautiful picture of bud-scales would be hard to find. The tough, leathery, hairy scales at the base of the cornflower's inflorescence once protected the delicate, folded parts of the undeveloped flower from wind and weather; now they do the same for the ovary and its seeds, while the flowerets wave their bright blue flags at passing bees.

Water

Plants require a number of basic elements in order to live, but of all of these, water is probably the most important, since it performs so many functions for the plant. From the beginning to the end of a plant's life, water plays a vital role, whether the plant be a desert species that subsists on the barest quantities of water, or a jungle plant that has had to adapt so as to keep from drowning in it.

When a plant germinates — or begins to grow — it is water that activates the seed, when its molecules permeate the tough seed coat and cause the embryo and endosperm inside to swell. A quantity of water equal to just 8 percent of the seed's total bulk is enough to awaken the embryo from dormancy. When it takes in more than 12 percent of its bulk in water a seed will start to grow, and from then on, its need for water will increase steadily as long as it lives. Water filling the cells and stretching tight their walls gives a plant rigidity, in the phenomenon known as *turgor*. Water absorbed by the root hairs carries dissolved minerals to the leaves of a plant; and from the leaves, water carries sugar, produced by photosynthesis, back down to the roots, where the energy stored in the sugar is used for further growth or stored for future use.

The system which plants have evolved in order to obtain water and to transport it throughout their structure is both ingenious and highly effective. Some aspects of this system defied rational explanation for years: how, for instance, does water rise to the uppermost branches of a 300-foot eucalyptus tree or a 100-foot Douglas fir? Many answers have been proposed for these and other questions, some of them still unsatisfactory; but the main parts of the story are now known.

The process by which water rises in a plant begins with *osmosis*, the passage of water through the semi-permeable cell wall in the root hair of a plant. Water passes through the cell wall because the cell contains a concentrated solution of sugars and salts, which can only be diluted, and brought into closer balance with water in the surrounding earth, by the entry of water into the cell. This fills the cell with water, stretching the cell wall until it is taut (or turgid, to use the botanist's term). The water then passes, by osmosis, to a neighboring cell, then to another neighboring cell, and so forth. The combined pressure of water in all of these cells results in the root pressure of the plant, which is the force that starts the water on its upward journey to the stem and the leaves.

Eventually, water entering the root of a plant reaches the vessels through which it will be conducted upward. These are located in that part of the root and stem known as the *xylem*. The xylem lies close to the middle of the root and stem structure, and is made up of cells that have died, but have left their walls in place, to create hollow, tubelike networks. There are several different kinds of these tubular networks, but their function is the same: they carry water gathered by the roots upward through the plant's structure.

The force that carries the water upward is still a matter of much learned discussion and research. For a while this force was thought to be the root pressure, but this theory collapsed when experiments showed that the root pressure was at its greatest in the early hours of the day, and decreased from then on. After this, it was thought that some sort of pumping action within the cells might be responsible for carrying water upward in plants, but this, too, has been disproved.

Today it is believed, with some reservation, that what carries water upward in plants is probably a phenomenon known as *mass streaming*. This occurs because there are basically two forces working on the water: osmosis in the form of root pressure, which in a sense pushes the water up from below, and *transpiration* in the leaves, which pulls water upward from above. If osmosis gains water for a plant, one might say that transpiration is the opposite process: it is the loss, by evaporation, of water into the open air from the leaves, and it is to replace this loss that more water moves into the leaf from the lower parts of the plant. The force of cohesion (the clinging together of molecules of water) also plays a role in moving water through a plant, and when it takes place in extremely narrow tubules such as those transporting the water in a plant, it has been shown to be a considerable force indeed.

Beans

These little mung beans are less than half the size of a child's fingernail, but they contain all that is required to make a complete new bean plant, and they are about to sprout.

Peeping through the widening rifts in the seed coats of these sprouting mung beans are the tips of the hypocotyls, the young roots which in later life will supply the plants with the minerals they need from the soil.

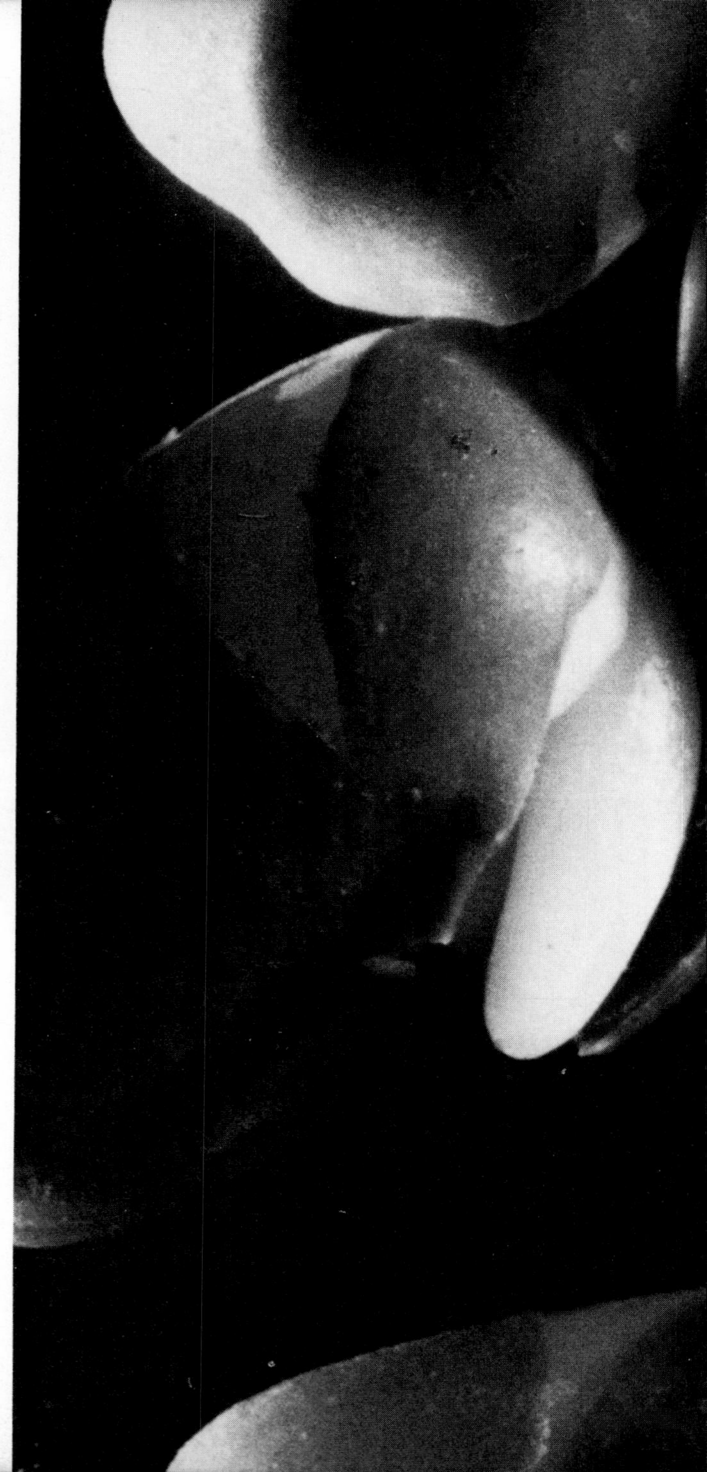

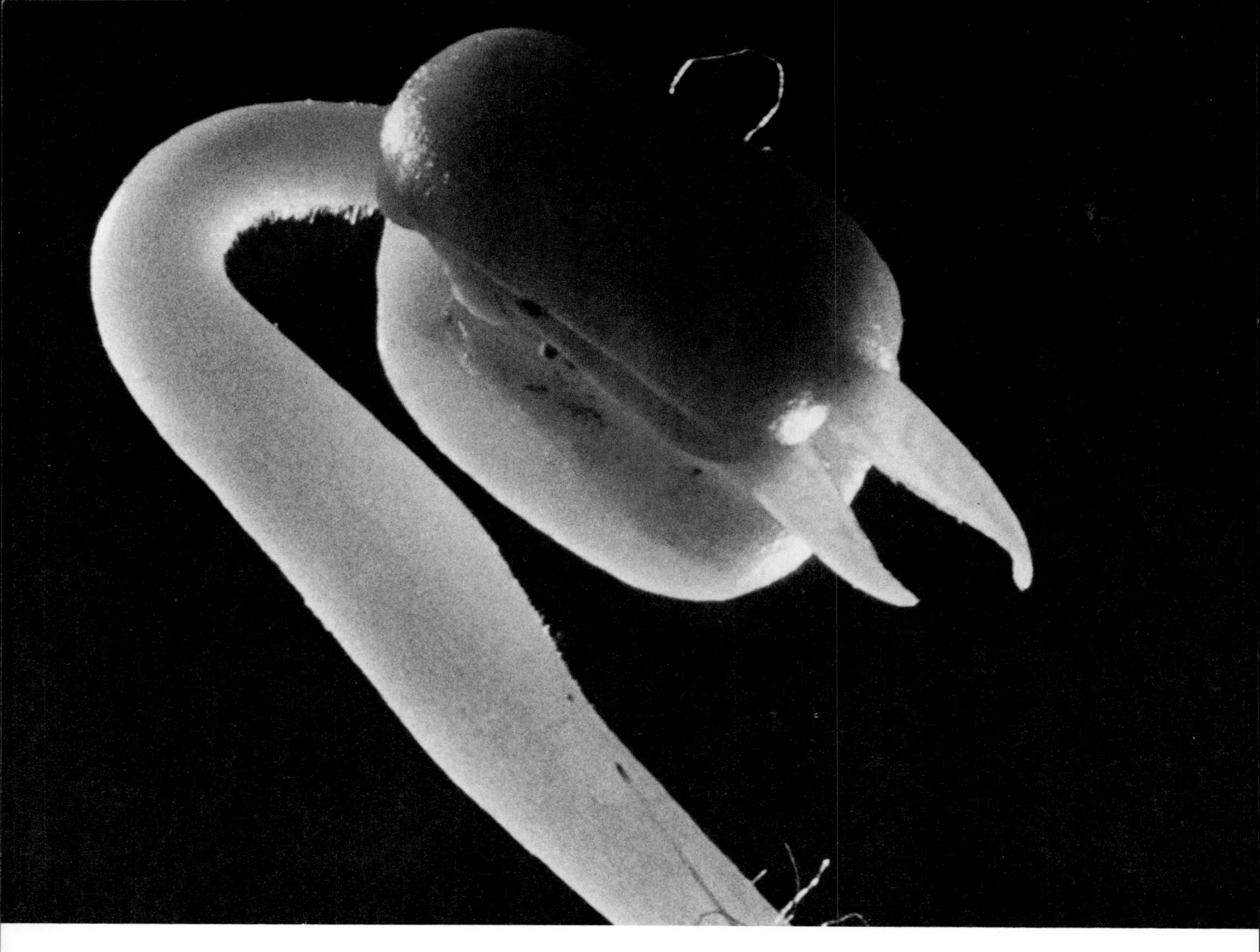

Here are the mung bean's cotyledons,
the nourishing endosperm without the seed
coat. The two pincer-like forms are the
first seedling leaves putting in their
appearance, while the long pale crook
is the root, already well-developed.

Still encased in the cotyledons,
the seedling leaves (already formed in
the embryo) are now enlarging
by cell elongation and are
being pushed outward ahead of the
apical meristem.

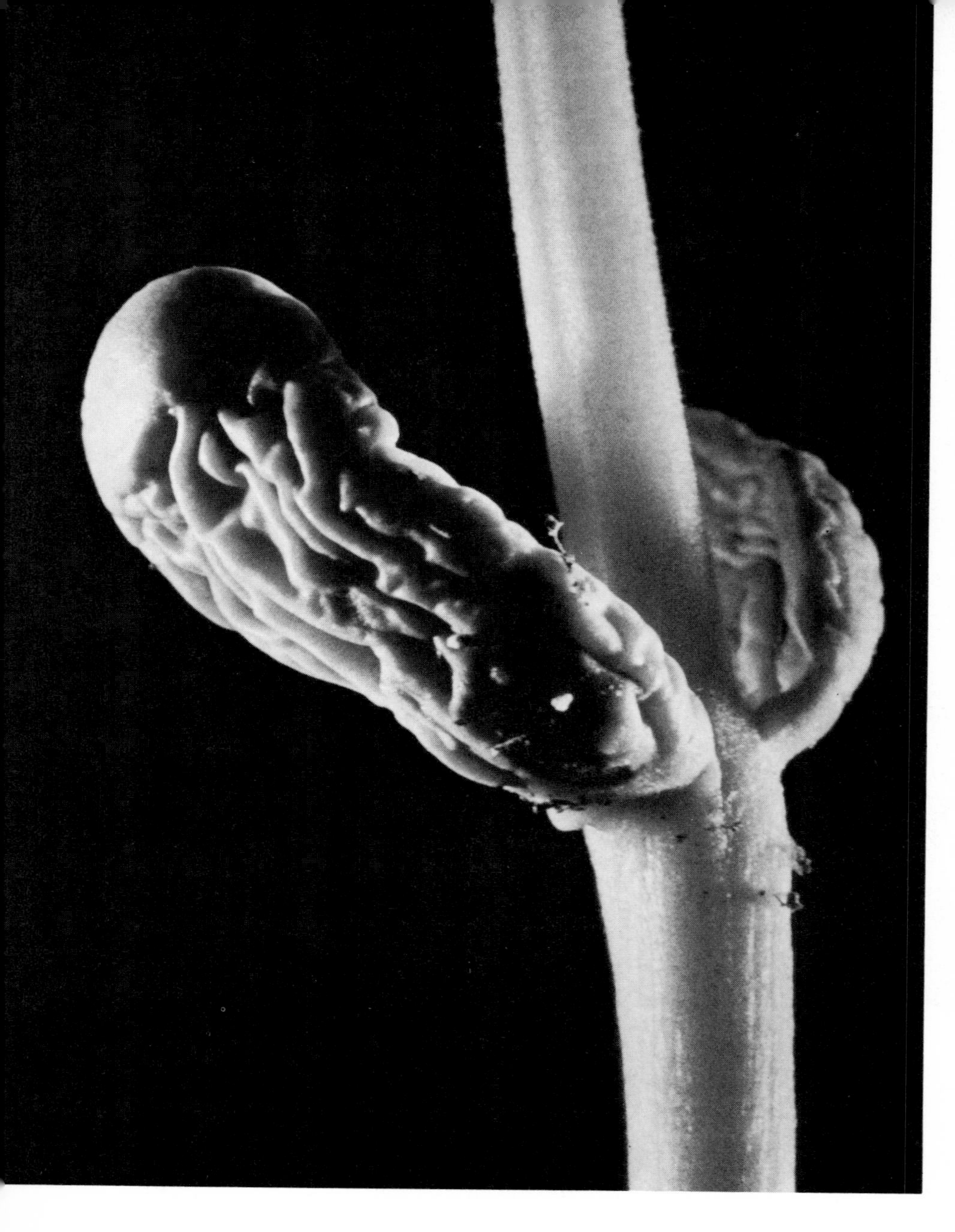

These shrivelled remnants on the proud, straight stalk of a bean plant are scarcely recognizable as the cotyledons that provided the infant shoot with its first nourishment. It is apparent that they have now given their all, so while the stalk grows, they will wither and eventually drop away.

Here is an excellent illustration of how seedling leaves develop in the small bean: the leaves have grown practically to their mature size without shedding the seed coat that originally enclosed the embryo. In other plants, leaf development often takes place later: the stem forms first, the leaves then branching out from it.

Look closely at the point from which the three small leaves branch: this is the apical meristem, the zone of cell division for new growth of the plant. The two big leaves, fully formed, will grow no more.

The bean's first flower raises its head to the sun, the giver of life and limitless energy. Below it more flowers, still tightly packaged in their casings, await the moment when they, too, will burst forth.

The birth of a green bean is not very different from the birth of such a dissimilar plant as a tomato. The slender bean is simply a ripening ovary, filled with fertilized seeds, as is the plump tomato or the fleshy green pepper.

Photoperiodism

One of the most famous — and beautiful — events of spring is cherry blossom time in Washington, D.C., when all of the Japanese cherry trees planted there long ago burst into bloom on a certain day. It is a wonderful sight indeed, when the branches laden with the delicate pink blossoms lend a magic color to the open areas around the great national monuments, to the parks, to many of Washington's streets, even to some backyards.

But have you ever stopped to think of how extraordinary it is that all of these trees, literally hundreds of them, in hundreds of different locations, may be in bloom on the same spring day?

That is what photoperiodism is all about — the response of plants to long days — such as in summer, or short — as in the springtime; or, to be more accurate, to short or long nights. For once photoperiodism had been discovered, with the Maryland Mammoth tobacco plant, the next conclusion was that it was not really the length of the light period that mattered to the plant, but the length of darkness.

This discovery came in 1938, from experiments on the cockleburr, a short-day plant which, in order to flower, must have no more than 15½ hours of light. The cockleburr was used because it is a hardy plant: it could, for instance, be stripped of its leaves, to see whether it was the leaves which determined the photoperiod. Surprisingly, the experiments showed that as little as one-eighth of a fully developed leaf left on the plant would promote flowering. With all of its leaves stripped away, however, the plant would not flower.

But the most important discovery made was that if, during the night, the plant was exposed to even a short burst of light — as little as one minute — it would not flower. It did not matter if the daytime hours were interrupted by short periods of darkness; what counted was an uninterrupted dark period.

This discovery led to others: it became apparent that the most effective light in preventing flowering of the cockleburr was red light of a certain wavelength. And the same light was equally effective in causing long-day plants — which required a fixed, uninterrupted light period — to flower.

An earlier experiment, with a certain kind of lettuce, had also explored the roles of light in plant growth. The lettuce, known as the Grand Rapids variety, germinated only when exposed to light; and the experiments had shown that red light most effectively stimulated germination. But if the red light that promoted germination was followed by red light of even a slightly different wavelength, so-called *far-red* light, the seeds did not germinate. Furthermore, after an entire series of red-light exposures, if the final exposure was to red light, the lettuce seeds would germinate; but if the final exposure was to far-red light, they would not.

The results of these lettuce experiments were applied to the cockleburr. What happened? A burst of red light at night prevented the cockleburr from flowering — but if this burst was followed by another burst, of far-red light, the effect of the red light was apparently cancelled, and the cockleburr flowered. By itself, however, far-red light did nothing. Apparently it functioned only as an agent that cancelled the effect of the red light.

The questions raised by photoperiodism are still a long way from being answered, but there is one thing that scientists do know now: that there is present in plants a pigment that exists in two forms, and that each of these two forms absorbs a different kind of red light. This pigment is called *phytochrome*. It is blue, and it will change color slightly when exposed to red or far-red light. One portion of the phytochrome molecule absorbs light, and is kin to a pigment found in red and blue-green algae; another, larger portion of the phytochrome molecule is protein.

What remains to be discovered is how phytochrome works in its triggering and inhibition of flowering. At this point it is believed to be associated with the biological clock — the keeper of the so-called circadian rhythms — which plants have in common with all other organisms.

Peppers

The ovary of a pepper plant, when it has ripened to a fruit, contains a huge number of seeds, sheltered by the thick, fleshy ovary walls.

This seed has ignored the inhibiting mechanism — which might be light, a hormone, or a moisture-temperature combination—which ordinarily prevents seeds from germinating prematurely. It has begun to form a baby pepper while still sheltered inside the ovary.

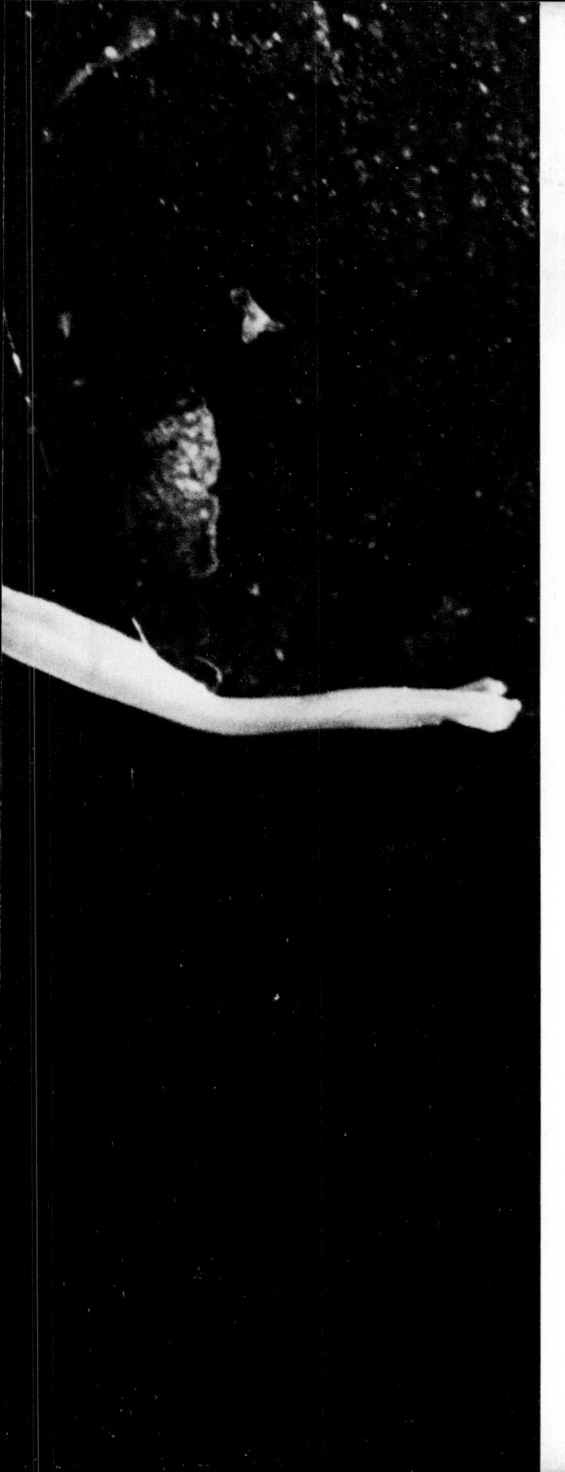

The cotyledons of this young pepper plant still bears the seed husks from which it came.

The long shoot reaching out is a pepper radicle, the first new bit of growth to appear. Its blind gropings have a goal; to reach the earth, where it will find nourishment for the plant.

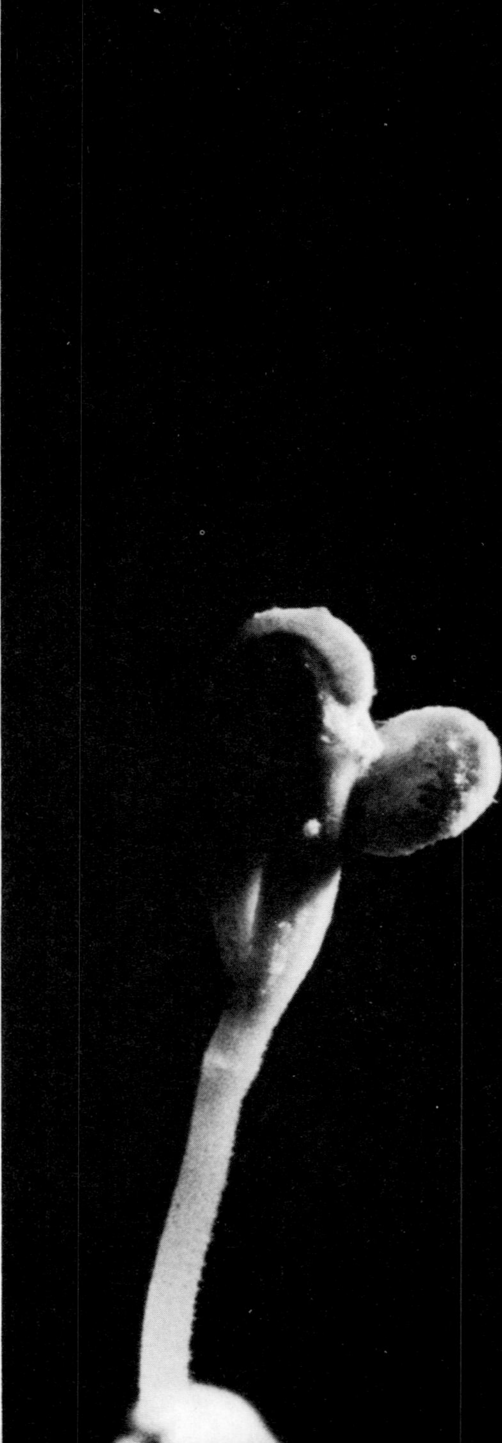

You will never see a better picture of cotyledons (above) than this one of the seedling leaves of a pepper plant still saucily wearing a seed husk for a hat. Each one of these shoots (right) will be forming functional leaves of its own which will grow from a point just below the cotyledons.

The graceful, limber little shoots of the pepper plant have grown into stems and branches (left), ridged for strength, ready to bear the weight of the fruit. But before the fruit appears we see the flower (right). The stamens — the reproductive organs rising in a ring around the central ovary — are ready to lose some pollen to a wandering insect or the wind. When the ovary itself has been pollinated, the fruit will begin to form.

The emerging fruit seems to be bursting forth from the final remnants of the flower. Dangling from its center is the withered stalk of the carpel, at which tip was the stigma, the sticky receptor for the pollen grains.

Color

When we think about plants, one of the first things we are likely to think of is color — and not just green, but the entire dazzling range of color we see in flowers, from pale pink through brilliant golds and reds to deepest purple. What makes these marvelous colors, and what is their purpose in plants?

Green — the basic color in all plants — is the color of chlorophyll, the pigment so prominent in photosynthesis. Because of this, it is safe to say that chlorophyll has probably played the most important role in the plant world for longer than any other pigment. Probably the first plants to develop chloroplasts — the tiny, sac-like organelles that contain chlorophyll — were the green algae, some 600 million years ago. From these algae, all other plants are believed to have originated. And for hundreds of millions of years thereafter, until the first flowering plants began to appear in the Cretaceous Period (about 135 million years ago), green was the dominant color in plants because photosynthesis was the primary plant function.

But then another function of color became important: pollination. The early plants, the gymnosperms, were all wind-pollinated, as many still are today — the conifers such as pine trees, for instance. The gymnosperms exuded a sticky sap at the proper season, and when the air was filled with wind-borne pollen some pollen grains would be caught by the sap, and the gymnosperm egg would then be fertilized. Gradually, however, in some way that we can only surmise, insects began to be involved in the pollination process, and with that the process itself began to change.

Insects were mobile, so that, in a sense, the plants could use them as transportation. Insects could carry pollen from flower to flower in a much more purposeful way than the wind allowed. Some kinds of insects were probably originally attracted by the sweet sap exuded to catch wind-borne pollen, and by the protein-rich pollen grains themselves. Some plants, through genetic accidents, developed adaptations favorable for attracting such insects: nectar, for instance, and then perhaps a certain bright color signaling to insects that there was nectar available.

In this way, insects and flowers influenced each other's evolution, and in due course an entirely new kind of plant appeared: the angiosperm, or flowering plant, whose seeds do not lie naked and exposed as do those of the wind-pollinated gymnosperms, but are protected within the fruits that develop from the ovules of a flower. And the color of a flower may tell some pollinating insects what it is, sometimes in so specific a way that only certain insects will come to certain flowers, bringing male pollen grains with them.

The color of a flower is affected by pigments other than chlorophyll. Chief among these pigments are the carotenoids, whose more than 60 varieties range from a light yellow to the deep orange-red of tomatoes. Carotenoids are also involved in photosynthesis — and there is some reason to think that they may have something to do with the phototropic, or light-responsive, movements of plants.

Finally, there are the anthocyanins, another large group of pigments which include reds and blues, from pale pink to the royal purple found in some varieties of pansies. The anthocyanins are dissolved in the cell sap of plants and are the pigments which are the least understood. They do explain color changes in some plants such as the morning glory, which starts the day pale pink and ends it pale blue: the reason the morning glory changes color is that anthocyanins react to acidity in cell sap, and in the course of a day the sap of the morning glory changes from slightly acid to slightly alkaline.

A spectacular display of plant pigments occurs in the fall in temperate North America, when the leaves change color from green to the dazzling golds and reds which, in New England particularly, so transform the countryside. These color changes take place because the production of chlorophyll decreases as cold weather approaches, permitting the yellow carotenoids, which have been present but masked by the deep green chlorophyll, to appear. At the same time, sugar produced by photosynthesis in the leaves is trapped by sudden physiological changes brought on by cold snaps, and this leads to the unmasking of the brilliant reds of the anthocyanins present in the cells.

Cantaloupe

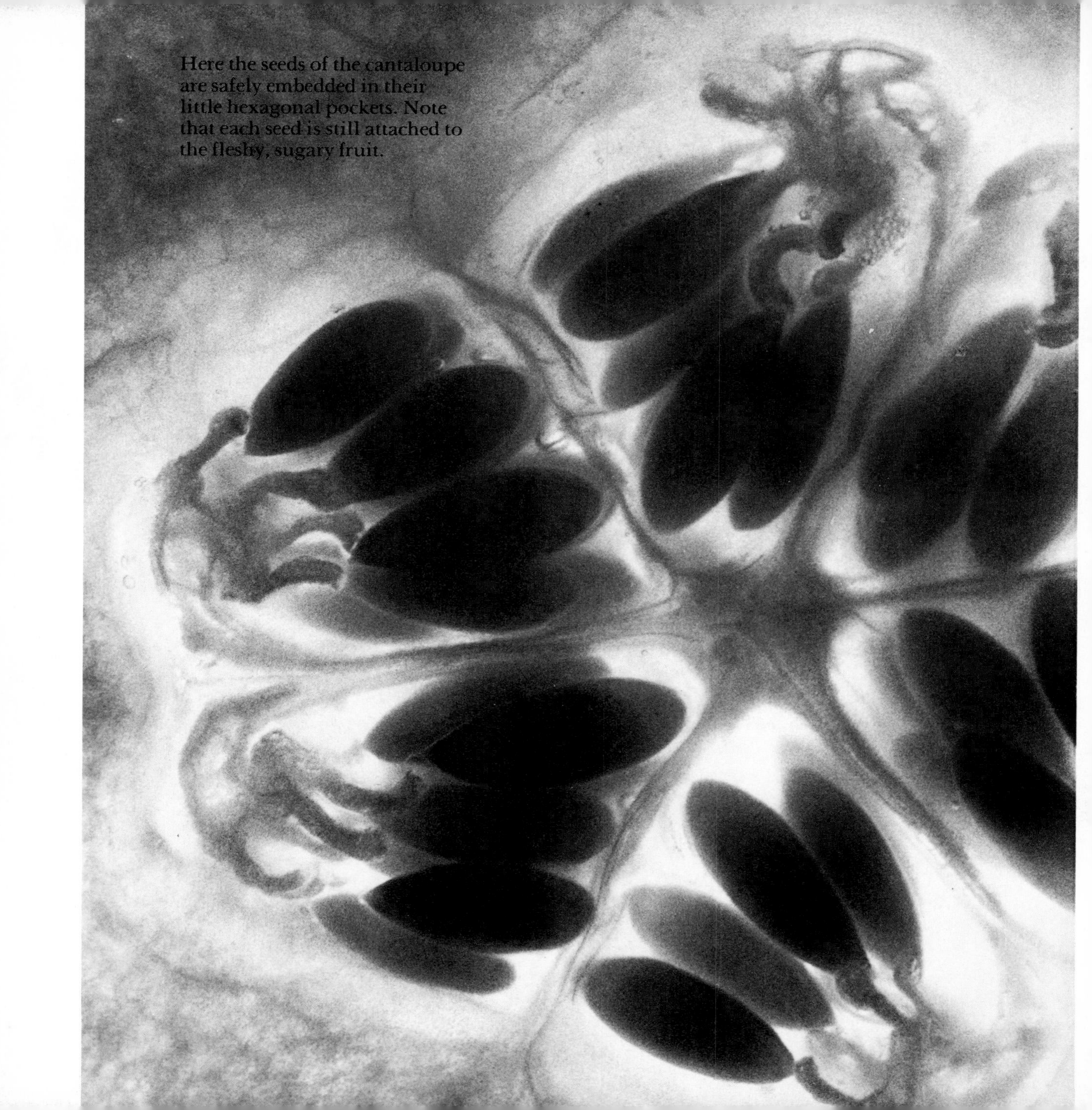

Here the seeds of the cantaloupe are safely embedded in their little hexagonal pockets. Note that each seed is still attached to the fleshy, sugary fruit.

The cantaloupe is among the most indiscriminate of plants, hybridizing so readily that it appears in numerous varieties. The seeds lie in the sugar-rich environment of the fruit or ovary, each bearing the potentiality for another plant.

Cantaloupe seedlings are literally pushed up through the ground, looking like shepherd's crooks. The pushing force is the hypocotyl which, in the other direction, drives the root tip through the soil. The bend of the "shepherd's crook" is called the epicotyl bend: as the seedling grows stronger, it will gradually straighten because of phototropism, the "light bending" force that keeps plants growing upward toward the sun.

This vigorous cantaloupe shoot still carries on its cotyledon the husk of the seed from which it grew.

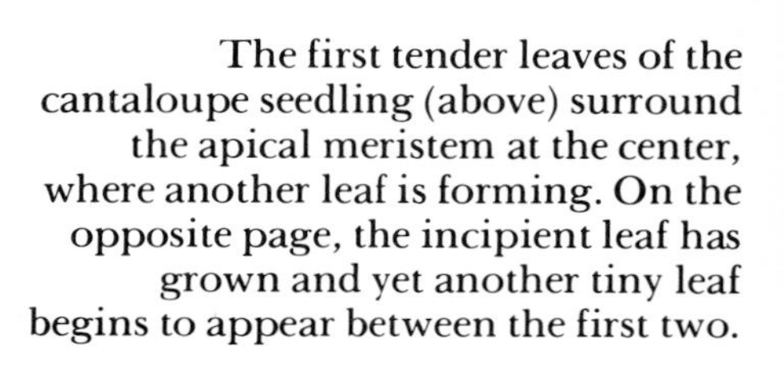

The first tender leaves of the cantaloupe seedling (above) surround the apical meristem at the center, where another leaf is forming. On the opposite page, the incipient leaf has grown and yet another tiny leaf begins to appear between the first two.

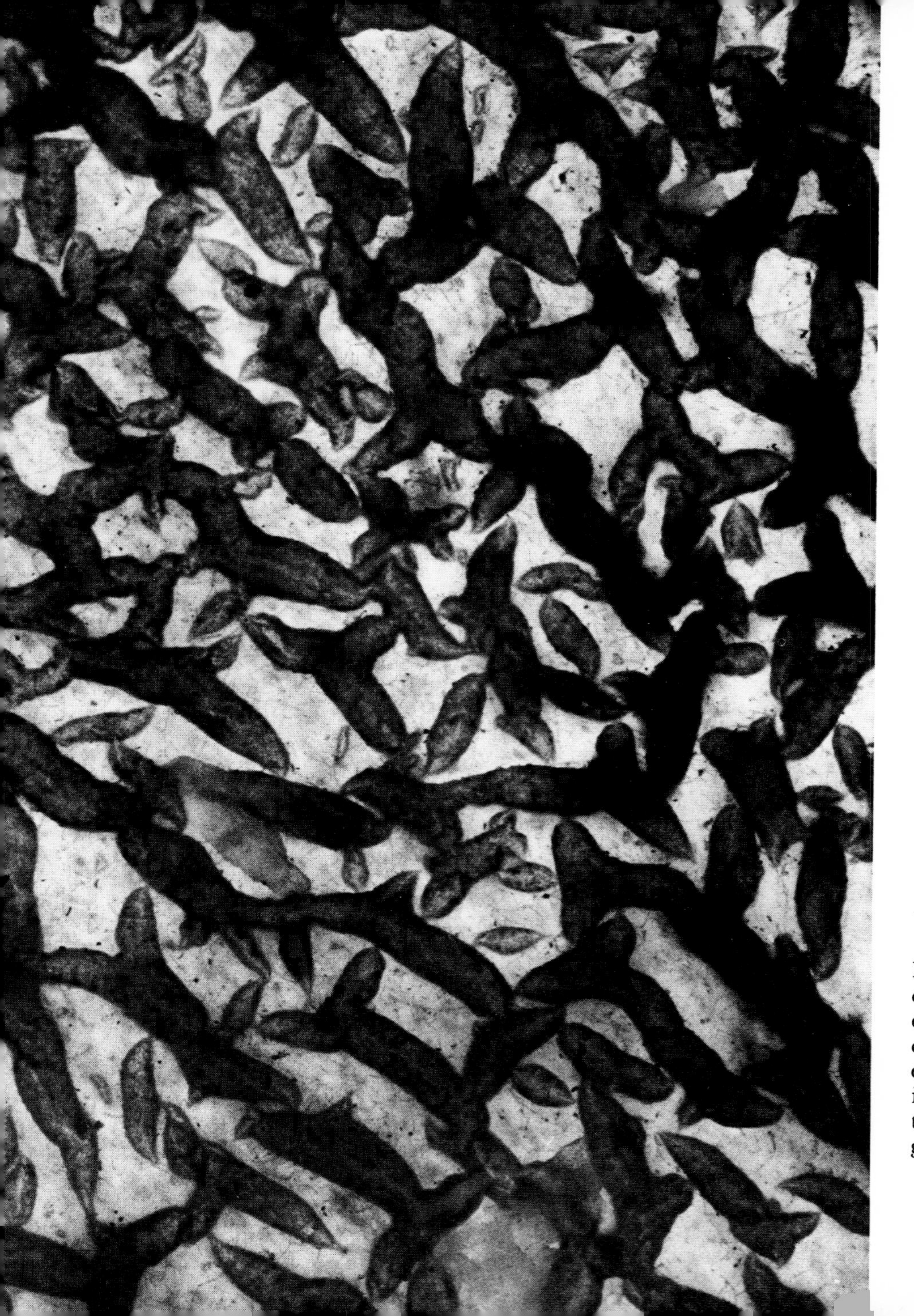

Notice the pattern of a cantaloupe's skin: thick, ridged, deeply sculpted and tough. This ovary must remain intact long enough for the fragile seeds in its succulent interior to develop to the point where they can germinate.

Tubers

There is a category of plants quite different from any other: the tuberous plants. A tuber is a grossly swollen part of the plant's stem, growing underground, which serves as a food storage area. Tuberous plants do not need to propagate from seeds, although they may do so; they can reproduce themselves directly from so-called buds on the tuber's surface.

Although several tuberous plants — including yams, sweet potatoes, and cassavas — are well known, the potato is the most familiar and is one of the food staples of the western world. The part of the potato plant which we eat is the tuber, and, as any cook knows, buds and sprouts will form on potatoes if they are kept in storage for a week or two. Outside, in the garden, these buds would form *stolons*, or runners, which would spread out in the soil and grow new tubers at certain places along their length and at their tips.

The size of a potato is determined not only by soil conditions but also by temperature. We tend to think of the potato as a hardy plant that will grow almost anywhere; actually, however, the potato has a rather narrow range of critical night temperatures within which it grows well. This range is around 35°F, and it explains why certain places such as Maine, Ireland, and Idaho are particularly good for growing potatoes, whereas others, like California, are not.

Besides having greatly enlarged stems and propagating themselves from buds, another way in which tuberous plants such as the potato differ from other plants is in the shape of the cells in the tuber. These cells have one purpose only: to store food — and they are shaped accordingly. Thus, while most plant cells are considerably longer than they are wide, reflecting the function of plant growth, the cells of a potato are about as wide as they are long, demonstrating their principal function as storage reservoirs for starch formed from the products of photosynthesis carried down to the storage tubers from the leaves.

Peanuts

Here the sprouting embryo of a
peanut is still sheltered by half
of the sturdy endosperm, which
is the part of the nut we eat.

The embryo of a peanut seed can be readily seen in the nuts we eat: it is the tiny, diamond-shaped point at the end where the two halves of the nut are joined. Here it puts forth tiny seedling leaves, one of the first signs of growth.

The peanut — half of which is shown in this picture — is still recognizable: the embryo is growing inside it, drawing its nourishment from the edible endosperm, which is also the food supply for the seed. From the back of the embryo a sturdy root tip has already come forth and now seeks to probe the soil for minerals and water.

The growing embryo splits the cotyledons of the nut in half.

The peanut at the right is shedding the papery husk or seed coat that covers its endosperm, which shrinks as it provides nourishment to the growing root (top). On the opposite page, both root and embryo are growing, the latter, as usual, from inside the two cotyledons. At first the root must do all the work, since it has to mine the soil for nutrient minerals and water for the new plant.

The first leaves of a peanut plant appear from between the cotyledons (below and right), an indication of advancing maturity, for where there are leaves photosynthesis can take place.

Folded tightly as they appear, the leaves gradually open (right) and will shed the now-shrunken cotyledons which harbored and sheltered them.

Beautifully regular and precisely aligned on the branch, these peanut leaves are poised to catch as much of the sun's light as they can. At night the leaves fold up — a measure common to plants of a tropical climate, which conserves their all-important store of water.

This lovely peanut blossom, once fertilized, will begin an unusual process in the development of its seeds. Groundnuts — as peanuts are often called — mature their seeds underground.

Already showing signs of withering, the peanut blossom, borne on its long stalk, begins to bend toward the earth. The developing ovary enclosed in the flower petals will burrow into the earth.

This maturing peanut (as it would appear dug from the soil) holds all the promise of a new plant. The homely pod also contains edible oils and proteins which are an essential food source to many millions of people.

About Percy Knauth

Those who know his other works may be surprised to find Percy Knauth writing the text for Esther Bubley's extraordinary photographs of growing plants, for while his interests, in a forty-year career in journalism, have been many and varied, botany has not been prominent among them. A foreign correspondent in pre-World War II times, private pilot, book and magazine editor and author, he has lately been best known for his work on mental depression, the outgrowth of a personal experience which led to his writing *A Season in Hell,* a book which won critical and professional acclaim in 1975.

There is solid professional knowledge behind his words on growing plants, however. Three years ago he edited a college textbook on the fundamentals of botany written by Professor Watson Laetsch of the University of California at Berkeley. This experience aroused a profound interest in the mystery and beauty of plant growth. When his wife Behri designed a book with Esther Bubley's photographs, Knauth found the challenge of writing it irresistible. The result is one of those rare products of collaboration: a true meeting of minds in a labor of love.

Appendix

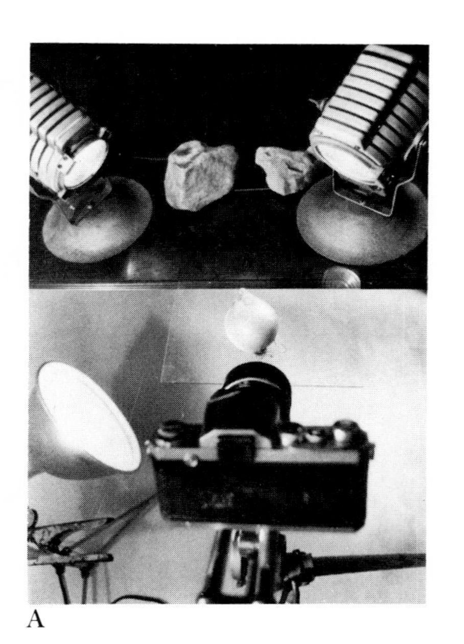

A

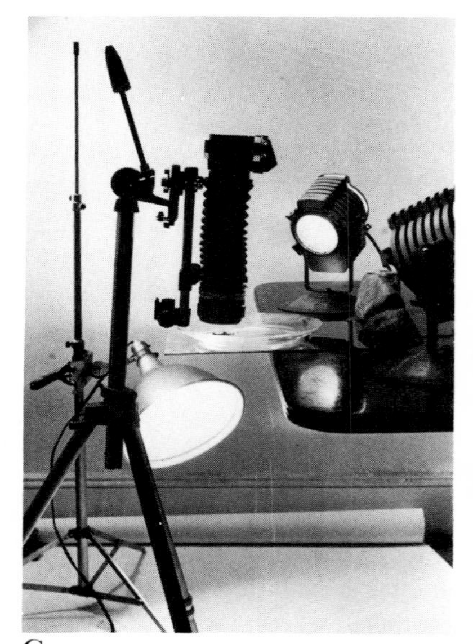

C

EQUIPMENT

A Nikon F Camera and 55mm. f/3.5 Micro Nikkor lens were used for these pictures. Nikon F Extension Rings and a Nikon Bellows 2 were used as needed, sometimes together for extremely close work.

Tri X Film was rated at a.s.a.600, exposed according to meter readings and developed in Acufine for 5½ minutes at 68 degrees. Except for the daylight shots, a sturdy tripod was used. Because of vibrations from the street, indoor exposures were 1/30 or 1/60, and apertures from 5.6 to f16.

B

LIGHTING

Two small spots — two photofloods
These are the basic lighting set-ups although variations were made according to subject matter and size.
Because of the heat of the lights, the sprouting seeds were photographed in water.

A Subject positioned on glass and lighted directly from below with floods. Spots sometimes used for accents.

B Spots falling directly on subject. Flood bounced from wall or ceiling for softening fill in.

C Subject positioned on glass. Floods directed at white paper background on floor, or in back of it. Spots sometimes used for highlighting.

D Daylight

Avocados

PAGE	LIGHTING	SHUTTER SPEED	APERTURE (f-Stop)
3	B	1/60	11
4-5	B	1/60	11
6-7	A	1/60	11
8	B	1/60	8
8-9	B	1/60	8
9	B	1/60	11
10	C	1/125	11
11	B	1/60	8
12	B	1/60	8
13	B	1/60	8
14	B	1/60	8
14-15	A	1/60	8
16	B	1/60	5.6

Beets & Radishes

PAGE	LIGHTING	SHUTTER SPEED	APERTURE (f-Stop)
19	B	1/60	11
20	A	1/30	5.6
21	C	1/30	11
22	C	1/30	6.3
23	B	1/30	6.3
24	B	1/30	8

Cucumbers

PAGE	LIGHTING	SHUTTER SPEED	APERTURE (f-Stop)
27	B	1/30	11
28	A	1/30	11
28-29	C	1/30	8
30-31	A	1/30	8
31	B	1/30	11
32-33	D	1/125	5.6
34	D	1/250	8
35	D	1/125	22
36	B	1/60	8

Cabbages

PAGE	LIGHTING	SHUTTER SPEED	APERTURE (f-Stop)
39	B	1/30	11
40	A	1/30	5.6
41	A	1/30	5.6
42	B	1/30	5.6
43	C	1/30	5.6
44	B	1/30	5.6

Peas

PAGE	LIGHTING	SHUTTER SPEED	APERTURE (f-Stop)
47	B	1/60	11
48Above	A	1/60	5.6
48Below	A	1/60	5.6
49	C	1/60	11
50	C	1/60	11
51	A	1/30	8
52	B	1/30	11
53	B	1/30	11
54-55	D	1/250	8
56	D	1/250	8

Potatoes

PAGE	LIGHTING	SHUTTER SPEED	APERTURE (f-Stop)
59	B	1/60	11
60	B	1/60	11
61	B	1/60	8
62	B	1/60	8
63	B	1/60	8
64	B	1/60	11
64-65	B	1/60	11
66-67	D	1/250	16
68	D	1/250	11

Tomatoes

PAGE	LIGHTING	SHUTTER SPEED	APERTURE (f-Stop)
71	B	1/30	11
72-73	A	1/30	5.6
74	A	1/60	11
74-75	A	1/30	5.6
75	B	1/60	8
76-77	B	1/30	5.6
78Above	B	1/30	8
78Below	B	1/30	8
79	B	1/30	8
80	D	1/250	8

African Violets

PAGE	LIGHTING	SHUTTER SPEED	APERTURE (f-Stop)
85	B	1/30	11
86-87	B	1/30	8
88	B	1/30	8
89	B	1/30	11
90	B	1/30	11
91	B	1/30	11
92	B	1/30	11

Onion

PAGE	LIGHTING	SHUTTER SPEED	APERTURE (f-Stop)
95	B	1/60	11
96	A	1/60	16
97	B	1/60	11
98Left	B	1/60	11
98Right	B	1/60	11
99	B	1/60	11
100-101	B	1/60	11
102-103	B	1/60	11
103	B	1/60	11
104	D	1/250	8

Squash

PAGE	LIGHTING	SHUTTER SPEED	APERTURE (f-Stop)
107	B	1/30	11
108	B	1/30	5.6
109	B	1/30	6.3
110Left	B	1/60	8
110Right	B	1/60	8
111	D	1/125	11
112	D	1/125	11
113	D	1/125	11
114	D	1/125	11
115	D	1/125	11
116	D	1/250	11

Cornflowers

PAGE	LIGHTING	SHUTTER SPEED	APERTURE (f-Stop)
119	A	1/30	8
120-121	C	1/60	5.6
122	C	1/60	5.6
123	C	1/60	5.6
124Above	C	1/60	8
124Below	A	1/60	5.6
125	A	1/60	5.6
126-127	A	1/60	11
128	B	1/60	11

Beans

PAGE	LIGHTING	SHUTTER SPEED	APERTURE (f-Stop)
131Above	B	1/30	8
131Below	D	1/250	11
132	C	1/30	8
133	C	1/30	8
134	C	1/30	8
135	C	1/30	8
136	B	1/30	11
137	B	1/30	11
138	B	1/60	11
139	D	1/250	11
140	D	1/250	8

Peppers

PAGE	LIGHTING	SHUTTER SPEED	APERTURE (f-Stop)
143	B	1/30	11
144	B	1/30	11
145	B	1/30	4
146-147	B	1/30	11
147	B	1/30	11
148	B	1/30	11
148-149	B	1/30	11
150	D	1/250	11
151	D	1/250	11
152	D	1/250	11

Cantaloupe

PAGE	LIGHTING	SHUTTER SPEED	APERTURE (f-Stop)
155	B	1/60	11
156	C	1/30	8
157	A	1/30	8
158-159	C	1/60	11
160	B	1/60	11
161	B	1/60	11
162	B	1/60	8
162-163	A	1/60	8
164	C	1/30	5.6

Peanuts

PAGE	LIGHTING	SHUTTER SPEED	APERTURE (f-Stop)
167	B	1/60	11
168-169	C	1/30	5.6
170Above	A	1/30	8
170Below	A	1/30	8
171	C	1/60	11
172	A	1/30	8
173	A	1/30	8
174Left	B	1/30	11
174Right	B	1/30	11
175	B	1/30	11
176Left	B	1/30	8
176Right	B	1/30	11
177	B	1/30	11
178	B	1/30	11